ON FRANK LLOYD WRIGHT'S CONCRETE ADOBE

*For their devotion to the development and administration of the
Architecture Archive and Architecture Museum
at the University of South Australia this book is dedicated to
Dr. Christine Garnaut, Dr. Julie Collins and Ann Riddle.*

On Frank Lloyd Wright's Concrete Adobe

Irving Gill, Rudolph Schindler and the American Southwe st

Donald Leslie Johnson
University of South Australia, Adelaide

LONDON AND NEW YORK

First published 2016 by Ashgate Publishing

2 Park Square, Milton Park, Abingdon, Oxon OX14 4RN
711 Third Avenue, New York, NY 10017, USA

Routledge is an imprint of the Taylor & Francis Group, an informa business

First issued in paperback 2016

Copyright © 2013 Donald Leslie Johnson

Donald Leslie Johnson has asserted his right under the Copyright, Designs and Patents Act, 1988, to be identified as the author of this work.

All rights reserved. No part of this book may be reprinted or reproduced or utilised in any form or by any electronic, mechanical, or other means, now known or hereafter invented, including photocopying and recording, or in any information storage or retrieval system, without permission in writing from the publishers.

Notice:
Product or corporate names may be trademarks or registered trademarks, and are used only for identification and explanation without intent to infringe.

British Library Cataloguing in Publication Data
Johnson, Donald Leslie.
   On Frank Lloyd Wright's concrete adobe : Irving Gill,
Rudolph Schindler and the American Southwest. -- (Ashgate
studies in architecture)
   1. Concrete masonry--California--Los Angeles--Case
studies. 2. Concrete masonry--History. 3. Wright, Frank
Lloyd, 1867-1959--Criticism and interpretation. 4. Gill,
Irving, 1870-1936--Influence. 5. Schindler, R. M. (Rudolph
M.), 1887-1953--Influence.
   I. Title II. Series
   721'.04454'092-dc23

ISBN 978-1-4094-2817-6 (hbk)
ISBN 978-1-138-24584-6 (pbk)

The Library of Congress has catalogued the print edition as follows:
Johnson, Donald Leslie.
   On Frank Lloyd Wright's concrete adobe : Irving Gill, Rudolph Schindler and the American Southwest / by Donald Leslie Johnson.
       pages cm. -- (Ashgate studies in architecture)
   Includes bibliographical references and index.
   ISBN 978-1-4094-2817-6 (hardback) 1. Wright, Frank Lloyd, 1867-1959--Criticism and interpretation. 2. Gill, Irving, 1870-1936--Criticism and interpretation. 3. Schindler, R. M. (Rudolph M.), 1887-1953--Criticism and interpretation. 4. Concrete houses--California--Los Angeles Metropolitan Area. 5. Concrete masonry. 6. Concrete blocks. I. Title.
   NA737.W7J65 2013
   721'.04450922--dc23
                                                                                                    2013003632

The letters, writings and drawings of Frank Lloyd Wright are copyright © 2013 The Frank Lloyd Wright Foundation, Scottsdale, Arizona: Wright Archive herein.

Contents

List of Illustrations — *vii*
Acknowledgements — *xv*
Frank Lloyd Wright Biography — *xx*
Preface — *xxi*
Introduction — *xxv*
List of Abbreviations — *xxvii*

1	Questions, Events . . . and Precast Concrete	1
2	The Buildings	31
3	The Taylors and the Griffins	53
4	Tiles and Blocks	73
5	Wright's Fiction	99
6	Historians' Fiction	119
7	Irving Gill, Regionalism and Concrete Adobe	143
8	Closure . . . Schindler and Resurgence	175

Appendix One — *199*
Appendix Two — *203*
References — *207*
About the Author — *217*
Index — *219*

List of Illustrations

0.1 Frank Lloyd Wright, November 1923   xix

1.1 Aline Barnsdall house on Olive Hill in the Los Feliz district of Hollywood, California, 1919-1922; called Hollyhock by Wright; aerial photograph of around 1924 looking northeast; major landscaping included new trees that more or less followed Lloyd Wright's suggestion; Residence A is northeast of the house, Residence B out of the picture to the left and near the road; the kindergarten was to be just off the picture and next to the road on the left; the vehicle entry court is north of the house stopped by a garage at its northern end; bedrooms are on the second level at the rear of the house; a future Municipal Art Gallery to right. Photograph held by Los Angeles Parks, from Levine, 1996.   4

1.2 Barnsdall house in Hollywood, California, 1919-1922; principal floor plan dated January 1921, revised August 1922, redrawn 1941; north is to the top of the drawing; note the plan components and spatial composition and relate to Figures 1.1, 7.5 and 7.7; bedrooms are in line on the second floor to the right; to the left of the courtyard is the loggia (the roof of which is the exterior stage), and to its left the living room with a reflecting pool outside; vehicle entry court is at the top with individual dog kennels in line to its right; landscape features not as built. From Wijdeveld, 1925, courtesy © 2013 The Frank Lloyd Wright Foundation.   5

1.3 Unity Temple in Oak Park, Illinois, 1905-1908; exterior view of the meeting room and at left the north entrance; exterior and some interior walls and the roof beams and slabs were constructed of poured *in situ* reinforced concrete; exterior walls finished with troweled surfaces evident in this photo but since removed; colonettes fronting high windows are precast concrete. Photograph probably by Henry Fuermann of ca.1908, from Wright, 1911.   15

1.4 Prototype houses at a Sewaren Improvement Company site in Sewaren, New Jersey, 1908-1909, Grosvenor Atterbury, Architect, extant?; an application of the Atterbury Standardized Sectional System he described as "fireproof, sanitary, and indestructible" and "intended to solve the problem of better living conditions for working classes at minimum cost;" project sponsored by Sage Foundation Homes Company; exterior walls textured off-form and unpainted. Contemporary photograph from Sloan, 1912.   20

1.5 "Concrete City" housing, Nanticoke Pennsylvania, 1913, Milton Dana Morrill, Architect; steel casement windows; reinforced concrete roof slab; wall plate (formwork) pattern untreated, only painted white inside and out with dark green trim. From Author's collection.   22

2.1 Alice Millard house in Los Angeles, California, 1923-1924; perspective drawing of 1923 by son Lloyd Wright and very much as built; exterior stairs located by Lloyd at the left of the living room leading from entry to lower garden; a large addition as a separate building was built to a design by Lloyd in 1925 and would be to the immediate left in this drawing and awkwardly butting a corner of the house. Courtesy and © 2013 The Frank Lloyd Wright Foundation.   33

2.2 Millard house, preliminary floor plan of entry or second level ("studio"), 1923; at left are two small sketches, one an elevation of a concrete tile and capstones and to their left a sectional detail entitled (at a later date) "textile block construction"; the building site ("plot") plan is sketched upper right; the garden elevation is similar to as-built; a preliminary side elevation is to the right. Drawing probably by Wright Junior. Courtesy and © 2013 The Frank Lloyd Wright Foundation.   34

2.3 Living room interior in 1980s, Alice Millard house in Pasadena, California, 1923-1924; furniture not Millard's. From author's collection.   36

2.4 John Storer house in Los Angeles, California, 1923-1924, photograph of ca.1924; colored canvas "awnings" above a roof terrace; in the background are houses that indicate steepness of the site; the front garden wall will be flush to a future sidewalk. From author's collection.   36

2.5 Above, Storer house, measured as built entry level plan with lower bedrooms at left; kitchen (with adjacent outdoor dining) and servants wing at right next to the garage; a swimming pool was added to the "rear terrace" in the 1980s. Below, Storer house, measured as built floor plans of the upper bedrooms at left with the living room and its roof terraces to their right; the upper loft floor plan shows access to roof terrace on top of bedrooms; these measured as-built drawings by Joseph R. Bateman of 1972 are similar to the construction plans. Both drawings courtesy Library of Congress, Prints & Photographs Division, HABS (CAL, 19-LOSAN-70-2,3).   37

2.6 Storer house, preliminary drawing of 1923 of the street or "South Elevation" with design thoughts sketched thereon and therefore not as built; high comb of "wood sash bars" with "leaded glass" inset; beams to support roof terraces' awnings; patterned garage doors at right not as built; block face pattern is that used for Millard house and not as-built for Storer; sixteen inches (406.4mm) module. Drawing probably by Wright Senior and Junior. Courtesy and © 2013 The Frank Lloyd Wright Foundation. 38

2.7 Samuel and Harriet Freeman house on a hillside above Hollywood, California, 1924; isometric drawing based on as-built measurements by Jeffrey B. Lentz in 1969. Courtesy Library of Congress, Prints & Photographs Division, HABS (CAL,19-LOSAN, 62-2). 38

2.8 Above, Freeman house, measured as-built entry floor plan; living room at top of drawing with a small roof terrace on top of closets to its left on; at right is the garage; squares in floors indicate precast concrete floor tiles; oak floor boards in living room; measured as-built drawings of 1969 by Robert C. Giebner "based in part on Wright's original drawing dated Jan., 1924". Below, Freeman house, measured as-built lower level floor plan; two bedrooms with large closets and terrace to the left, one bedroom wall was removed in late 1920s; square precast concrete floor tiles; to the right a storage area under the garage was converted to a small bed-sit guest apartment to a Schindler design. Measured as-built drawing by Jeffry B. Lentz of 1969 "based in part on Wright's original drawings dated Jan., 1924." Both courtesy Library of Congress, Prints & Photographs Division, HABS (CAL,19-LOSAN, 62-3, 4). 40

2.9 Living room interior in 1960s, Freeman house in Los Angeles Hills overlooking downtown Hollywood. From author's collection. 41

2.10 Charles and Mabel Ennis house in Los Angeles, California, 1924-25; view of the south elevation bathed in setting sun light displays the bulky mass of the dining room above the great retaining wall; each functional area is articulated by a different pattern of concrete tiles and by volume size; large exterior planting boxes as part of the retaining wall are below tall windows to dining area. Photograph by Balthazar Korab, courtesy Christian Korab. 42

2.11 Ennis house, main floor plan with garage and servants' quarters to the left of the entry "courtyard"; car and foot entrance is under a foot bridge that connects servants (at left) to the main house; in the 1940s the then owner John Nesbitt installed a swimming pool in the north-side garden parallel with the long corridor; the retaining wall is at the bottom edge of the platform. Courtesy and © The Frank Lloyd Wright Foundation. 44

2.12 Kindergarten on Olive Hill in Hollywood, California, 1923, for Aline Barnsdall; only foundation walls were constructed; perspective from the west; a wading pool is to the left; entry at top left; the raised central roof above the single classroom was to sit on oddly formed and exposed scissor trusses; this the only extant perspective drawing (now in poor condition) of the final design probably executed by Wright Junior. Courtesy and © 2013 The Frank Lloyd Wright Foundation. 45

2.13   Plan parti of the Millard, Freeman, and Storer houses, and Barnsdall kindergarten, no scale. Drawing by Author. 46

4.1   Plan and elevation of their own one room cottage (named Pholiota) by architects Walter and Marion Griffin located in Heidelberg near Melbourne, Victoria, built 1920 to 1921 of knitlock wall construction and roof tiles; after construction drawings. Drawing by Paul Stark for Johnson, 1977. 75

4.2a/b Knitlock construction system, patent drawings of various wall segments in plan and elevation as submitted by the Griffins and David Charles Jenkins (assignee) as published in 1917; roof tiles (not shown) patented 1918. From Navaretti, 1998. 77-78

4.3   Blueprint of a concrete block and tile patent drawing of 1923 that was not lodged, "Inventor Frank Lloyd Wright"; figures 6 and 7 are blocks as used for the Millard house; figures 1-5 are a double wall system as devised for the Storer house although the face design shown is for Millard; steel wires run vertically and horizontally in edge grooves; figure 4 shows a solid core wall, figure 5 a hollow core. Courtesy and © 2013 The Frank Lloyd Wright Foundation. 82

4.4   Evolution and comparison of Griffin's concrete knitlock tiles (A) with Wright's blocks for the Millard house (B) and a similar design (D), and Wright's textile blocks (C and E), 1916 to 1924. Drawings by Author to the same scale. 84

4.5   Drawing of four relief patterns on the exterior faces of Wright's textile blocks; A for Storer and Ennis but first applied and in various sizes to the German Warehouse of 1915+; B for Freeman and the solid square left of center was now and then replaced with clear glass; C for Storer as is D but the designs were derived from precast elements prepared in terra cotta first for the Larkin Administration Building (1903-1905), then the German warehouse (1915+), then for the Tokyo Hotel during 1919-1922; one half of a symmetrically split A and C were used occasionally to clad posts or as wall coping, i.e. the top course of a wall. Based on archival material, drawing by Author. 86

4.6   Two-dimensional "Line-ideas" composed within a square as drawn by Arthur Wesley Dow; all drawings are variations of pattern No. 17, top left, except the middle three at the top and the bottom line that show samples of "dynamic" compositions. A page yellowed by age from Dow, 1899. 89

4.7 Wright's textile tile system as drawn by Richard Neutra ("RN") in 1925; prepared for and published in his article for de Fries, 1926; the tile face pattern design is for the Ennis house; the drawing was later modified by Wright's office using English words and metric measurements, and modified again using English words and measurements, each often published. Courtesy and © 2013 The Frank Lloyd Wright Foundation. 91

4.8 Four concrete block shapes from among hundreds available commercially throughout America in 1919; compare E (which is Nel-Stone) with C. Drawing by author based on information supplied by the Portland Cement Association and as found in Sweeney, 1994. 93

5.1 Monolith Homes commissioned by Thomas P. Hardy as a prototype for a series of houses at an unknown site in Racine, Wisconsin; project designed and drawn by Rudolph Schindler July 1919; floors, walls, and roof to have been poured *in situ* (monolithic) concrete. Courtesy and © 2013 The Frank Lloyd Wright Foundation. 101

5.2 Pictorial explanation of "double shell" wall construction as applied to the Imperial Hotel in Tokyo (1917-1923); as it appeared in *Liberty* magazine for the general public's edification. 109

5.3 Midway Gardens in Chicago, Illinois, 1913-14, a view of the north belvedere in 1914; two designs of concrete wall tiles (glued to brick walls) and two for pressed metal fascia; wood molds used for precast concrete wall tiles, miscellaneous spires and spindles and for Alfonso Innelli's sculptures; demolished 1929. Photograph by Henry Fuermann and published first in 1915. 111

5.4 "Proposed Women's Building, county fair grounds" near Spring Green, Wisconsin, a preliminary design of 1914; plan and perspective of entry with mothers' rest room at left; drawing probably Herbert Fritz, Sr. From Alofsin, 1993. 111

6.1 Exterior view (above) and ground level plan (below) of the "Governor's House" or "The Palace" at the Mayan temple site of Uxmal in Yucatan, Mexico. View a portion of foldout frontis piece to vol. 1, Stephens (1843). 124

6.2 Presentation drawing of a project for a "Science Hall future chemistry building" at the University of New Mexico in Albuquerque, dated "14 Dec. 1915", Walter and Marion Griffin, Architects; building scale is deceptive--note entry doors; face of the pronounced lintel course to have been ornamented presumably with precast terra cotta elements; not as built 1916-1917; faint lines of pencil on tracing paper. Courtesy National Library of Australia, Eric Nicholls Collection. 128

6.3 Owls' Club in Tucson, Arizona, Henry Trost, Architect, of ca.1903, photograph of about 1904; building refurbished in 1987. Courtesy of Arizona Historical Society/Tucson (41445). 129

6.4 Warren Hickox house in Kankakee, Illinois, 1899-1900; perspective drawing based on a photograph. Prepared for and published in Wright, 1910. 130

6.5 A.D. German warehouse in Richland Center, Wisconsin, 1915-20; "Front Elevation"; blueprint of a construction drawing; third level exterior attic described as "made of concrete stone;" a two storey "studio annex" for crafts people attached at left; poles east and west corners decorative only Courtesy and © 2013 The Frank Lloyd Wright Foundation. 131

6.6 Frederick C. Bogk house in Milwaukee, Wisconsin, 1916; preliminary design of exterior attic decoration (described as a "stone lintel" but acting as a frieze) presumably to have been in precast concrete and colored terra cotta; not as built; colored pencil, ink and water color. Courtesy Library of Congress, Prints and Photographs Division. 133

6.7 Described as a "Portion of Exterior of Nahuan Palace" when published in Viollet-le-Duc (1876); note awning and compare to Storer house; note the stone upper string courses and integrated ornament; this conjectural reconstruction based on the west building of the Quadrilateral Court of the Nunnery at the Mayan temple site of Uxmal in Yucatan, Mexico. 135

7.1 Nelson E. Barker house in San Diego, California, Irving Gill, Architect, 1911-1912; presentation drawing of "Front Elevation" by Gill; close to as built except for garden walls; masonry walls stuccoed on the exterior, plastered on interior; elevation may appear symmetrical but is not. The only ornament in Gill's post-1907 houses was finely detailed and oiled wood cabinetry and stairways in the interiors. Courtesy San Diego Historical Society. 148

7.2 A "cube house" project by Irving J. Gill, Architect, of ca.1910, perspective drawing by Lloyd Wright; Gill's first cube house was for G.W. Simmons in San Diego of 1909; all walls masonry with stucco on the exterior and plaster on interior. From Gill, 1916. 150

7.3 House project for G.P. Lowes at Eagle Rock, California; perspective drawing by Wright Junior and dated 1922. From Wright (1925). Courtesy and © 2013 The Frank Lloyd Wright Foundation. 153

7.4 George and Alice Millard house in Highland Park, Illinois, 1906; perspective drawing; wood stud and joist construction; exterior entirely of glass and wood board-and-batten siding stained with ferric oxide; a classic house among those that helped define the Wright School. Prepared for and published in Wright, 1910. 154

LIST OF ILLUSTRATIONS   xiii

7.5   Barnsdall house in Hollywood, California, 1919-1921; theater stage with clay tile floor on loggia roof; courtyard steps lead down to loggia; steps up to stage and pedestrian access to each roof; the ornamental theme of abstract hollyhocks is found on pinnacles and on oversized column capitals on the north side of the court. Photograph by and courtesy of Neil Levine.   152

7.6   Horatio West Court in Santa Monica, California, Irving Gill, Architect, 1919; isometric view of 1960 drawn by Robert C. Giebner; internal functions articulated; masonry walls coated off-white on interior and exterior; windows steel framed. Courtesy Library of Congress, Prints & Photographs Division, HABS (CAL,19-SANMO,1-2).   156

7.7   T.P. Martin house for Taos in New Mexico, R.M. Schindler, Architect, a project of 1915 and immediately after Schindler's trip through the American southwest; floor plan drawn by Schindler; mention of "adobe constr," i.e. sun dried mud brick walls, not concrete blocks; stables at top; dine at bottom left and snooker table room symmetrically each side of the living room before an exterior pool; kitchen in the left wing; bedrooms in right wing; the plan predates Wright's similarly planned Barnsdall house of 1919. Courtesy Art Museum, University of California, Santa Barbara.   166

7.8   Martin house, perspective drawing by Schindler of internal "court" looking toward living room; note uneven adobe wall surfaces coated with stucco. Courtesy Art Museum, University of California, Santa Barbara.   167

8.1   Architect R.M. Schindler's own Kings Road House in West Hollywood; north to top; first floor plan dated "Nov. 1921" and very close to as built 1921-1922; Chase apartment spaces top right and Schindler apartment bottom, each with exterior courtyard/garden; room names on drawings comply with building application; guest room middle left with its garden; all rooms with openable sliding glass walls to gardens; faint lines of colored pencil on tracing paper by Schindler. Courtesy Art Museum, University of California, Santa Barbara.   176

8.2   Schindler's Kings Road House "aerial isometric" of 1969, drawing by Jeffrey B. Lentz based on measured drawings; Schindler studio bottom right; tilt-up concrete wall panels with glass in vertical space between each; floors of exposed concrete painted and polished; two second floor sleepouts here enclosed. Courtesy Library of Concress, Prints & Photographs Division, HABS (CAL, 19-LOSAN, 68-2)   177

8.3   El Pueblo Ribera Court in La Jolla, California, R.M. Schindler, Architect, 1923-1924; presentation drawing; working module of four feet (1.22M) so the living room is twelve feet deep (3.65M) with an openable window wall to the patio similar in concept to his Kings Road house; the bottom right drawings of "form [work] details" and "structural details" help explain the slip form system perhaps based on observing Lloyd Wright's; top left is a view down on the "roof terrace" trellis; stairs to the roof are not enclosed; concrete was left exposed and unpainted; wood trellis and trim were darkly stained; owners often transformed their roof terrace into an enclosed bedroom. Courtesy Art Museum, University of California, Santa Barbara. 178

8.4 El Pueblo Ribera Court in La Jolla, 1923-1924; photograph ca.1943; upper "roof terrace" enclosed for bedroom; circulation to it and bedroom to right at ground level through courtyard; scribed joints between concrete pours; walls untreated. Courtesy Library of Congress, Prints & Photographs Division, HABS (CA.1943-5). 179

8.5 Philip Lovell house in Los Angeles, Richard Neutra, Architect, 1927-29; final design of 1928; structural frame made of standard steel products; concrete retaining walls; facades of steel, glass, and panels of gunite-applied concrete; entry at top right; an open air swimming pool is at the lowest level straddled by the house; photograph of 1929. Luckhaus Studios, from the Collection of Thomas S. Hines. 183

8.6 Dorothy M. Foster house for an unknown site in Buffalo, New York, a project of 1923; garden elevation; exterior bearing walls probably brick; concrete or clay tile blocks appear to be ornamental; sixteen inch (406.4mm) design module. From Fisher Fine Art, 1985. Courtesy and © 2013 The Frank Lloyd Wright Foundation. 187

8.7 Larkin Company Administration Building in Buffalo, New York, 1903-05, demolished 1950; steel frame structure (protected by fired terra cotta) with some brick bearing walls; reinforced concrete floor slab and beam construction; perspective drawing of the street facade based on a photograph. Prepared for and published in Wright, 1910. 188

Acknowledgements

An academic cannot satisfactorily conduct historical research without considerable help, not least that offered by many writers and scholars whose dissertations and published works have been a source of knowledge and inspiration. Many are identified in the references and notes herein. Additionally, it is with great pleasure and gratitude to acknowledge the generosity, encouragement or assistance of the following people and organizations. Each in unique or special ways assisted in this project.

In Scottsdale: Bruce Brooks Pfeiffer, Director, Indira Berndtson, Administrator Historic Studies, and Oskar Munoz, Assistant Director, the Frank Lloyd Wright Foundation, and their dedicated staff. In Tucson: the architects and professors Duane Cotè and the late Kirby Lockard. In the Los Angeles area: Eric Lloyd Wright (son of Lloyd and grandson of Frank); historian Kathryn Smith; curator and historian Robert L. Sweeney; guide Jock de Swart; August O. Brown (eighth owner of the Ennis house); volunteer staff at Barnsdall house; Joel Silver (in the 1980s owner and restorer of the Storer house); Prof. Thomas Hines at the University of California; and Prof. Jeffrey M. Chusid when curator of the Freeman house.

In Santa Barbara: the late Prof. David Gebhard and staff of the Art Museum of University of California; Dr. Gerald and Marian Groff; and Eréne Groff.

In Seattle: staff at the College of the Built Environment at the University of Washington, my alma mater and where later I was a visiting research fellow, and especially Betty Wagner, now retired Librarian and my boss back in 1955-1957.

In Australia, Adelaide: Prof. Mads Gaardboe, Prof. Donald Langmead, Dr. Christine Garnaut and Dr. Julie Collins at the University of South Australia. In Melbourne: Prof. Miles Lewis and Jeff Turnbull at Melbourne University, and Peter Navaretti as Heritage Strategy Planner at RMIT University. In Sydney: Prof. Robert Freestone and Dr. Bronwyn Hanna at the University of New South Wales; Dr. Anna Rubbo at Sydney University. Late of Champaign, Canberra and Brisbane, in Perth, Prof. Christopher Vernon at the University of Western Australia.

In Orlando, Florida, Adam Kaare Johnson for his work on Florida Southern College. In Rockledge, Florida, Emeritus Prof. Paul E. Sprague. In Richland Center,

Wisconsin, the late Margaret Helen Scott. In Champaign/ Urbana Prof. Paul Kruty at the University of Illinois. In Skokie, Illinois, research and publications staff of the Portland Cement Association.

The owners and caretakers of the many buildings personally visited over past decades were patient and generous with knowledge, time and coffee.

Research assistance was gratefully received and facilitated by a travel grant from the U.S. National Endowment for the Humanities; by a series of grants from the Australian Research Committee; by financial assistance granted on a number of occasions by colleagues on the research committees at Flinders University and the University of South Australia; by encouragement and assistance from Bruce Brooks Pfeiffer, Indira Berndtson and Oskar Munoz at the Frank Lloyd Wright Foundation Archives in Scottsdale, Arizona; by Brent Sverdloff and Pamela Kratochvill and other staff at the Getty Center for the History of Art and the Humanities in Santa Monica; and by staff at the following institutions and places: Northwest Architecture Archives, William Gray Purcell Papers at University of Minnesota (herein Purcell papers); University of California at Santa Barbara, Art and Architecture Collections (herein SB Archives); art or general librarians and/or museum staff at Flinders University (especially the diligent interlibrary loans [ILL] people); University of South Australia; Adelaide University; State Library of South Australia; National Library of Australia; Mitchell Library and the State Library of New South Wales, Sydney; Library of Congress, Washington, D.C.; University of California at Santa Barbara; University of Washington in Seattle; Burnham and Ryerson Libraries at the Art Institute of Chicago; Avery Art and Architecture Library in Columbia University, New York; and the Research Library, Special Collections of the University of California at Los Angeles for the Lloyd Wright Collection.

The University of South Australia and Flinders University, both in Adelaide, continue to generously support my research in many significant ways including grants.

The papers of George and Florence Taylor held by the Mitchell Library, Sydney, whose staff were most helpful, do not contain material of the period prior to 1920 or about Walter and Marion Griffin or the Wrights. Many relevant drawings and other documents were sighted at or received from the Wright Archive in Scottsdale, the Burnham Library, the Avery Library, Library of Congress, National Library of Australia, and Australian National Archives.

Very capable research assistants during the early stages of research were Mrs. Elizabeth Beck, Alan Johnson now in South Africa, and Christine Garnaut (née Smerdon) now director of the Architecture Museum at the University of South Australia.

Quotations of Wright's letters and certain published work are copyright and courtesy of the Frank Lloyd Wright Foundation. Quotations of Lloyd Wright's correspondence are courtesy of Eric Lloyd Wright and the University of California at Los Angeles. A special thank you to the people and organizations that granted permission to reproduce illustrations and use quotations. They are identified in the illustrations' captions and elsewhere as appropriate.

Wright and Griffin scholars who have read portions large and small of the manuscript in progress were: Prof. Anthony Alofsin at the University of Texas, Austin;

Prof. Robert Freestone at the University of the New South Wales; Prof. Thomas Hines at the University of California Los Angeles; Dr. Christine Garnaut and the late Prof. Donald Langmead at the University of South Australia; Prof. Miles Lewis at the University of Melbourne; Emeritus Prof. Mati Maldre at Chicago State University; Emeritus Prof. Paul E. Sprague of the University of Wisconsin-Milwaukee; Dr. Anna Rubbo at Sydney University; Prof. Christopher Vernon, Landscape Architecture at the University of Western Australia. Publication and technical staff of the Portland Cement Association read applicable sections.

Encouragement and valuable comments were obtained from readers for the *Art Bulletin* in New York, *Fabrications, Journal of the Society of Architectural Historians Australia and New Zealand*, and *Winterthur Portfolio* in Delaware. A reader for the *Journal of the Society of Architectural Historians*, Chicago, offered useless comments on an early draft.

The content of this book, however, remains my creation.

A special thank you to Series Editor Eamonn Canniffe and to Valerie Rose, Gillian Steadman and the Production staff at Ashgate Publishing.

To Sonya Hasselberg, wife and weaver extraordinaire, and to our sons Karl and Adam, thank you for all that is worthwhile.

Ashgate Studies in Architecture Series

SERIES EDITOR: EAMONN CANNIFFE, MANCHESTER SCHOOL OF ARCHITECTURE,
MANCHESTER METROPOLITAN UNIVERSITY, UK

The discipline of Architecture is undergoing subtle transformation as design awareness permeates our visually dominated culture. Technological change, the search for sustainability and debates around the value of place and meaning of the architectural gesture are aspects which will affect the cities we inhabit. This series seeks to address such topics, both theoretically and in practice, through the publication of high quality original research, written and visual.

Other titles in this series

Architect Knows Best
Environmental Determinism in Architecture Culture from 1956 to the Present
Simon Richards
ISBN 978 1 4094 3922 6

Nationalism and Architecture
Edited by Raymond Quek and Darren Deane, with Sarah Butler
ISBN 978 1 4094 3385 9

The Political Unconscious of Architecture
Re-opening Jameson's Narrative
Nadir Lahiji
ISBN 978 1 4094 2639 4

Forthcoming titles in this series

The Dissolution of Place
Architecture, Identity, and the Body
Shelton Waldrep
ISBN 978 1 4094 1768 2

Building Apartheid
On Architecture and Order in Imperial Cape Town
Nicholas Coetzer
ISBN 978 1 4094 4604 0

Fig. 0.1   Frank Lloyd Wright, November 1923
Source: As published in the June 1925 issue of the Amsterdam architectural journal *Wendingen*

Frank Lloyd Wright Biography

In 1867 Frank Lincoln Wright was born in the rural town of Richland Center, Wisconsin, to a Baptist minister and musician, William, and a public school teaching assistant, Anna. William's religious, artistic and ethical beliefs and actions were crucial to Wright's maturation. Anna's farming family, the Joneses, were followers of Emersonian and Unitarian philosophy that had a profound and lasting effect on Frank. Early formal education in Wisconsin and Massachusetts was followed by a few technical classes at the University of Wisconsin in Madison.

Changing his name to Frank Lloyd Wright, he was employed in Chicago as a factotem and draftsman by architect Joseph Lyman Silsbee during 1886 and 1887, and then as a draftsman by architects Dankmar Adler and Louis Sullivan from 1887 to 1893 when, as a result of a crippling national financial crisis he and other employees were laid off. Wright immediately established a private architectural practice in Chicago, a home office in nearby Oak Park, Illinois (1893-1911), and then another he called Taliesin outside Madison, Wisconsin, on family farm lands near the rural village of Spring Green (1911-1959). He operated temporary offices in Chicago (1911-1924), Tokyo (1917-1922), and Los Angeles (1919-1924). Wright was in a formal partnership with Chicago architect Webster Tomlinson during 1901-1902 and thereafter acted in association with but few architects. A second or winter home-office called Taliesin West was designed for ancient desert lands near Scottsdale, Arizona: construction began in 1937 and continues.

In 1914 his lover, Mrs Mamah Cheney, her children and three employees were murdered at the Wright home in Spring Green. After his wife Catherine divorced him in 1922 he married Miriam Noel, who suffered morphine addiction and they finally divorced in 1927. Olgivanna Hizenberg became his third wife in 1928. They formed an apprenticeship program in 1932 called the Taliesin Fellowship. It carries on today as a school of architecture within a wider professional practice.

Wright received national and international awards, degrees, and other high honors including Gold Medals from the Royal Institute of British Architects in 1941 and the American Institute of Architects in 1949. He died in 1959.

All illustrations are of Wright's buildings and designs unless noted otherwise.

Preface

The case study presented in this book is of Frank Lloyd Wright's design of four concrete block houses of 1923-1925, of the Barnsdall house of 1919-1921 and secondarily a proposed kindergarten for it. The five residences are extant, if distressed or half destroyed by earthquakes and rebuilt at enormous expense, each situated in the metropolitan Los Angeles area. Of necessity, this book is also about the irregular historical speculations and discordant attributions and analyses published about the buildings during the past eighty-five years. The concrete blocks and their houses have been subjected to words and phrases similar to these examples: the blocks were uniquely moulded; construction was woven like a textile fabric; steel rod reinforcing formed a spider's web; the blocks were a groundbreaking or a modern or an unprecedented construction process; the process was cheaper than standard construction; the blocks an original masonry; they presaged a ground-breaking or a modern or a new architecture; they were an integration with nature; derived from Egyptian, Mayan or maybe Aztec architecture; part of the "modern movement"; that the Barnsdall house (said to be a copy of an ancient Mayan temple) was a prototype; and on and on. Apart from often using architectural and constructional terms incorrectly, such comments were, we can generously say, usually unwitting. Yet they were, and are, all too common. That the unlearned repeated those comments is understandable simply because their sources were the trusted learned whose fashionably imaginative history was accepted. This book straightens a deformed record.

The record also indicates that too often buildings are presented to wondering or adoring audiences as objects resident on the edge of cultural emptiness. Not mentioned are the client's desires or explanations, or the creator's biographical and concurrent professional and social imperatives. Additionally, opinions and assurances can be prejudiced by an uncritical acceptance of an architect's own propaganda and by the utterances of cloying followers. Now and then, only visual things appended to a building inside and out are surveyed; architectonics ignored. When gathered together such prejudices act as a catalyst to devolution into false myths soon trapped in popular history. Those presentations are especially noticeable in many if not most of the voluminous publications by and about Wright (1867-1959) and his building and other designs from 1886 to 1959. By isolating the four unique concrete block buildings from Wright's oeuvre, each commissioned at a calamitous moment in his life, we can record typical deceptions and provide enlightenment.

The methodology employed is straight forward. Before beginning these essays the buildings and their environments had been visited and experientially studied. The first essay presents a useful history of concrete as a structural material for buildings. Two methods of application are then introduced: Grosvenor Atterbury's prefabricated slabs followed by Milton Dana Morrill's *in situ* formwork, and for each a resulting aesthetic. Then by weighing new evidence against old and re-evaluating texts contemporary with design and construction, by looking at concrete technology, construction methods, the designs, architectonics, and aesthetical conditions and attributes, the concrete blocks and houses are delineated and analyzed contextually. They are compared to a series of fictional, fanciful chronicles by Wright and to wayward historical accounts and critical evaluations by others, then and more recent. The explanations herein are managed by attention to what was said, when it was said and why, and of its factual accuracy, yea, possible, or nay. As always, biographical and architectural revelations that correspond to reality are far more curiously interesting, more provocatively revealing than sterile popularized fictions.

Therefore, the exotic, regional and historical determinants of the five buildings are analyzed by outlining contemporary building practices in concrete construction; by offering correct chronologies, critically examining Wright's publicly uttered words and related social and professional actions before and during the period; and by discovering the relevant interrelationships of his architect son Lloyd Wright (FLW Junior), of two other Southern California architects, Irving Gill and Rudolph Schindler, and of the American/Australian architects Walter and Marion Griffin, both once employed by Wright and later designers of Canberra, capital of Australia.

This is followed by essays not theoretical but practical. They discover why Wright began to design concrete blocks and where he might have gotten ideas for them. The physical and aesthetic properties of his blocks are then studied. This examination occupies chapters 2 through 5. and includes suspicious fictional accounts offered by Wright and other writers who emulated and/or wrote uncarefully about what we refer to here as his Concrete Adobe.

Chapter 6 places this information and Wright's buildings in the theoretical milieu as practiced then. Included are reasonable thoughts about the designers' reactions to the historical and geographical Southwest region. As research progressed it became clear that Gill was not merely a regionalist, as often proposed, but the region inspired a prescient universal theory of architecture. I argue that that inspiration was absorbed by Wright who then integrated it into his own romantically-driven theoretical musings. They in turn were remodeled and refreshed in the 1930s, a process discussed in my book *Frank Lloyd Wright versus America: the 1930s*, and less obviously in a sequel entitled *The Fountainheads: Wright, Rand, the FBI, Hollywood*.

The closing chapter looks at the disenchantment beset upon those proud, talented people as their design productivity succumbed to societal and aesthetic conservatism. It then promotes the work of Gill and of former Wright employee Schindler as catalytic to a resurgent modernism; a universal modernism but without Wright's now-changing yet idiosyncratic aesthetical manners.

The creation of architecture is a design and construction process. For a building's users it is an experiential activity, vision the most important sense. It follows that when one analyses a building, illustrations are essential to help explain the process and something of the experience. But this book is a study in architectural history, not a picture book. Presentation drawings communicate how a building will appear while construction drawings describe how it will be put together, how building materials and components are fitted one to another. Drawings that may be difficult for some readers are easy for the architect, construction manager, and architectural historian.

The buildings under study have undergone some alterations or restoration during the past ninety or so years. Perhaps most obvious are additions to the Millard house. Original gardens indicate something of an architect's intentions or those of the commissioning client, or both. All landscapes have changed with fashion.

This is a study in architectural theory achieved by applying the persuasion of history. It therefore enlivens discussions on the methodology of architectural history. As well, it assumes the reader understands that architecture is a design profession, that buildings are created by persons for people within society, that architectural theory, if potent, is also apparent during the creative process and in the experiences of a building's users: Architecture is the most socially relevant of design endeavors.

The buildings discussed herein were created--designed--in response to the complexities of a client's program. By the architect's abilities they gain certain aesthetic values that we prize more highly than, say, a slap-dash bicycle shed. Since each was designed to be useful they are, therefore, architecture, not art. As Oscar Wilde has neatly informs us, "All art is quite useless", without material function. Do we denigrate design disciplines when we refer to them or their useful works by the honorific term "art"?, that is (or was) a synonym for "craft". Yet, blessed by ignorance it pleases us to do so.

Introduction

In the 1920s Frank Lloyd Wright (1867-1959) claimed to have invented a new system of concrete block construction and aruged its superiority. Since then people attempted to uphold the claim adding it presaged a new architecture and would satisfy the demand for cheap housing. Further, many people have argued that Wright borrowed architectural elements from ancient buildings and applied them to his own. To determine the correctness of those assertions we need to present Wright's declarations and the support of others, and then to test their accuracy and veracity. Evidence is found in technical developments for concrete blocks before Wright's sad experiences in Los Angeles began in 1919 when he designed a manor house and four concrete block houses 1923-1925. It is revealed experientially in the buildings, in the manner of his design process, in the reading of relevant pictorial and textual documents, and by a study of the extant buildings. In this process we also discover the full impact upon Wright of his contemporaries, the architects Irving Gill in San Diego and Rudolph Schindler in Los Angeles. Moreover, the independent advancements of those two architects renewed the promise of a viable modernism in the 1920s. However it was not Wright's technology or aesthetics. Thus a clarifying interpretation of intent is assembled and placed in historical, biographical and theoretical contexts. Only then is it possible to fully appreciate Wright's, Gills' and Schindler's buildings beyond their architectonic and experiential qualities.

List of Abbreviations

Author (date), for example Sprague (1990), refers to an author and the publication date as set out in References that follow the Notes.

AForum. *Architectural Forum*, New York City.

ARecord. *Architectural Record*, New York City.

Burnham. Burnham and Ryerson Libraries, Art Institute of Chicago.

Fabrications. *Fabrications. Journal of the Society of Architectural Historians, Australia and New Zealand.*

FLW. Frank Lloyd Wright.

FLWQ. *FLW Quarterly*, FLW Foundation, Scottsdale, Arizona.

HABS. Historic American Buildings Survey, Prints & Photographs Division, Library of Congress, Washington, D.C.

JSAH. *Journal of the Society of Architectural Historians*, Chicago.

LW Papers. Lloyd Wright Papers, Special Collections, Research Library, University of California at Los Angeles.

PArch. *Progressive Architecture*, New York City.

Purcell papers. William Gray Purcell Papers, Northwest Architecture Archives and Manuscript Division, University Libraries, University of Minnesota.

SBArchive. Architecture Archives, Art Museum, University of California, Santa Barbara.

TP&H. *Town Planning and Housing Review*, Sydney.

WArchitect. *The Western Architect*, Minneapolis/Chicago.

WHistory. *Wisconsin Magazine of History*, Madison.

Wright Archive. Material and letters personally sighted in or received from the Frank Lloyd Wright Archive, Scottsdale, Arizona, or as received from the Archives of the History of Art, Getty Center for the History of Art and the Humanities, Santa Monica, California.

1

Questions, Events . . . and Precast Concrete

**Reason and Will have been exalted by Philosophy and Science.
Let us now do homage to Imagination.**
— Frank Lloyd Wright, 1927

Before a building appeals to the intellect it will have appealed to the senses.
— John Burchard and Albert Bush-Brown, 1961

In the creative individual's mind, more often than not there is no precise line between invention and borrowing, between originality and appropriation. The unconscious subliminal aspect of human creativity is not easily measured by either the activator, producer, participant, or observer. During the architectural design process (the most important conceptual activity in imagining a future building), an awareness of appropriation can be blurred by mental fevers in devising, contriving, discovering, even puzzling. They can be subverted by hidden mental dishonesties induced by ignorance, fantasy, unconscious acts, and self-imposition. Receiving and applying inspiration is inherently a personal riddle.

Borrowing can be holus bolus or particular, acknowledged (in music, for example, variations on a theme), or not. But when not explicitly admitted in words, its identification can be problematic, accusation withheld. In Frank Lloyd Wright's case some recent research has proven valuable in illuminating the extent of the problem through clarification and insight. Perhaps the first and certainly one of most useful was research conducted for an exhibition and documentary film that was published in an extensive, well-illustrated catalogue, all with the general title of *Frank Lloyd Wright and Madison: Eight decades of artistic and social interaction*. Under the editorship and guidance of architecture historian Paul E. Sprague, the researchers he marshaled relied on primary documentation and interviews, not on hearsay or propaganda from any source, or hagiography, and presented a series of essays about the various architectural commissions. As well, the authors were counseled to avoid "drawing general conclusions about Wright's methods, attitude or visual strategies". Sprague described the catalogue as Wright's "designs and executed buildings in and near Madison, Wisconsin, [and] records the artist's life and work there from 1879, when his family settled in the Madison area, until his burial in the Lloyd-Jones cemetery near Taliesin in 1959".

Donald Hoffman's well documented study of Wright's Fallingwater house for Edgar Kaufmann—basically a concrete structure—near Mill Run, Pennsylvania was first published in 1978 and then generally ignored for a couple of decades. Finally it was revised and republished in 1993. His book about Hollyhock house and another about the Dana House of 1902-04 were equally dedicated to objectivity.[1] More recently is Franklin Toker's compendious frank, humorous, and useful *Fallingwater Rising* of 2003.

Anthony Alofsin's erudite essays about possible European and other precedents for Wright's architectural and figural ornamentation prior to around 1920 provided ample proof of the multiplicity of probable sources of particular appropriations, subtle or blatant. As well, we hasten to add, Alofsin acknowledged Wright's own creations. Joseph Siry's documentation and dense analyses in a book about Unity Temple set aside myth, clarified the client's program, sorted design evolution, explained relevant concrete construction techniques 1900-1908, and suggested influences and precedents for that extraordinary building. Two of Siry's latest contributions to architectural history are of reinforced concrete structures for Wright's Imperial Hotel in Tokyo and for the Pfeiffer Chapel at Florida Southern College.

Also useful to the context of these essays is Kathryn Smith's history of the Aline Barnsdall commissions given to Wright for buildings in Chicago and Los Angeles during 1915-1931.[2] The information Robert Sweeney offered in the compact book *Wright in Hollywood* about Wright's concrete block buildings for Barnsdall contained a narrow band of historical detail while interpretations were linear and traditional. Kevin Nute's informative research into possible Japanese models as sources for Wright's two and three dimensional designs was less objective or comparative (to Western or Chicago traditions) but otherwise a valuable contribution. Paul Kruty's study of Midway Gardens presented new information and detail on that highly influential concrete and tile building of 1913-14. And there was my own book, *Frank Lloyd Wright versus America: the 1930s* published in 1990.

While these illustrated books, and others, for the most part ignore or treat too lightly the subject of Original Action, as do *all* biographies or quasi-biographies of Wright, they do acknowledge, as had Wright, the guiding inspiration of, for instance, preindustrialized cultures, the nineteenth century French architect and theorist E.-E. Viollet-le-Duc, England's interrelated Arts and Crafts Movement as it devolved in America, Viennese secession of the 1890s, and traditional Japanese two-dimensional art.[3] Yet, Wright and some of his students, followers, observers and assenters (hagiographers in the main) have maintained that he was the author of many inventions. One is the focus of this study.

What concerns us here is not the advancement of claims that Wright went outside his own mind and borrowed things or ideas created elsewhere. All artists reluctantly admit verbally, or by exposition, their eclecticism, if not its degree or the propriety of its elements. Rather, we want to know what prompted those supposed appropriations, their magnitude, their intervention during Wright's design process, and therefore about the role of invention. Particularly, do Wright's claims or those of his assenting observers hold firm under scrutiny.

Back in 1831 the prescient author Mary Shelly suggested that to invent was an organic process, intuitive rather than linear or rational. "Invention [she said] consists in the capacity of seizing on the capabilities of a subject; and in the power of moulding and fashioning ideas suggested to it".[4] This was probably close to Wright's understanding. We shall carry on more practically, keeping in mind the thought of Arthur Conan Doyle: "What one man can invent another can discover". For our purpose, to invent is to originate as a product of one's own contrivance something useful that did not exist before. (The words "something useful" also admits a potential legal constraint.) It is uneasily distinguished from acts of emulation, simulation, mimicry, imitation, even parody, but is persuaded by assimilation, absorption, and other acts induced by influence and inspiration, even at the subliminal level.

Our investigation will study five houses in the Los Angeles area designed by Wright and built between 1919 and 1925 under his often indirect supervision or under that of his first-born son or that of his trusted draftsman Rudolph Schindler. Cement stucco applied to fired hollow clay blocks were the major construction materials for the walls of one now called Hollyhock. It was designed and built for the oil heiress, brazen political progressive, arts patron and theater director and producer, Aline Barnsdall, from June 1919 and into 1921. Some people have referred to Barnsdall as the ultimate iconoclast, a "pinko", communist, anarchist, or the socialist she was. After twenty-four years of trailing Barnsdall the Federal Bureau of Investigation concluded she was a harmless, bohemian-styled lady of wealth.[5] Wright found her "neither neo, quasi, nor pseudo," disarming and disingenuous, and as "domestic as a shooting star". She was also a confidant of another Wright client in Los Angles, the Freemans and also of the Lovells who were clients of architects Rudolph Schindler and Richard Neutra. The title "Hollyhock," apparently the lady's favorite flower, was not Barnsdall's but given by Wright first as late as 1932 in his autobiography. Until then it was known as her residence or Barnsdall house.[6] Her stucco-faced mansion set within sloping bush land, olive trees in broken orchard lines, bound by streets and suburbia, still sits proudly centered on top of a high round hill on the edge of Hollywood in Los Feliz (see Figs 1.1 and 1.2). Back then it had views of surrounding citrus orchards and west to the great ocean. Now it sits in city-owned Barnsdall Art Park next door to an imposing municipal art gallery and within a treed landscape constructed post-1950.

Patterned precast concrete blocks were employed for the walls of the other four more modest houses of 1923 into 1925. Each of the five is heritage listed and known worldwide. That adulation, however, is based on an enjoyment of them as works of design, yes, but also on suspect—often erroneous—information that these essays will endeavor to expose, politely, and correct.

Fig. 1.1 Aline Barnsdall house in the Los Feliz district of Hollywood, California, 1919-1922

The architectonics and appearance of the four houses, Wright often told us, were derived from technological, historical, regional *and* universal considerations. Surely the client's needs were also influential. There was a moment when he also said that his concrete block system of construction for the four houses was his invention.[7] In other instances people have said he borrowed from—at times implying that he copied—ancient buildings. (Perhaps most famously was historian Vincent Scully's list of 23 "borrowings" based mainly on visual comparison.[8]) As a corollary, therefore, we need to discover the evidence in support of Wright's various declarations and, of course, that in support of assertions made by observers, art and architectural historians in the main.

We need to test the veracity of claims, including the possibility that feigned, imagined or made-up stories were presented as firm evidence. Answers, or at least explanations, are found in technical developments of concrete block before Wright's sad experiences with Los Angelenos began in 1919. They are found in the manner of his design process, and a careful reading of relevant pictorial and textual documents and of extant buildings. Only then can a clarifying interpretation of intent be assembled and placed in historical, biographical and theoretical contexts. Only then can we fully appreciate the buildings beyond their visual qualities. The revelations are—the history is—not as expected.

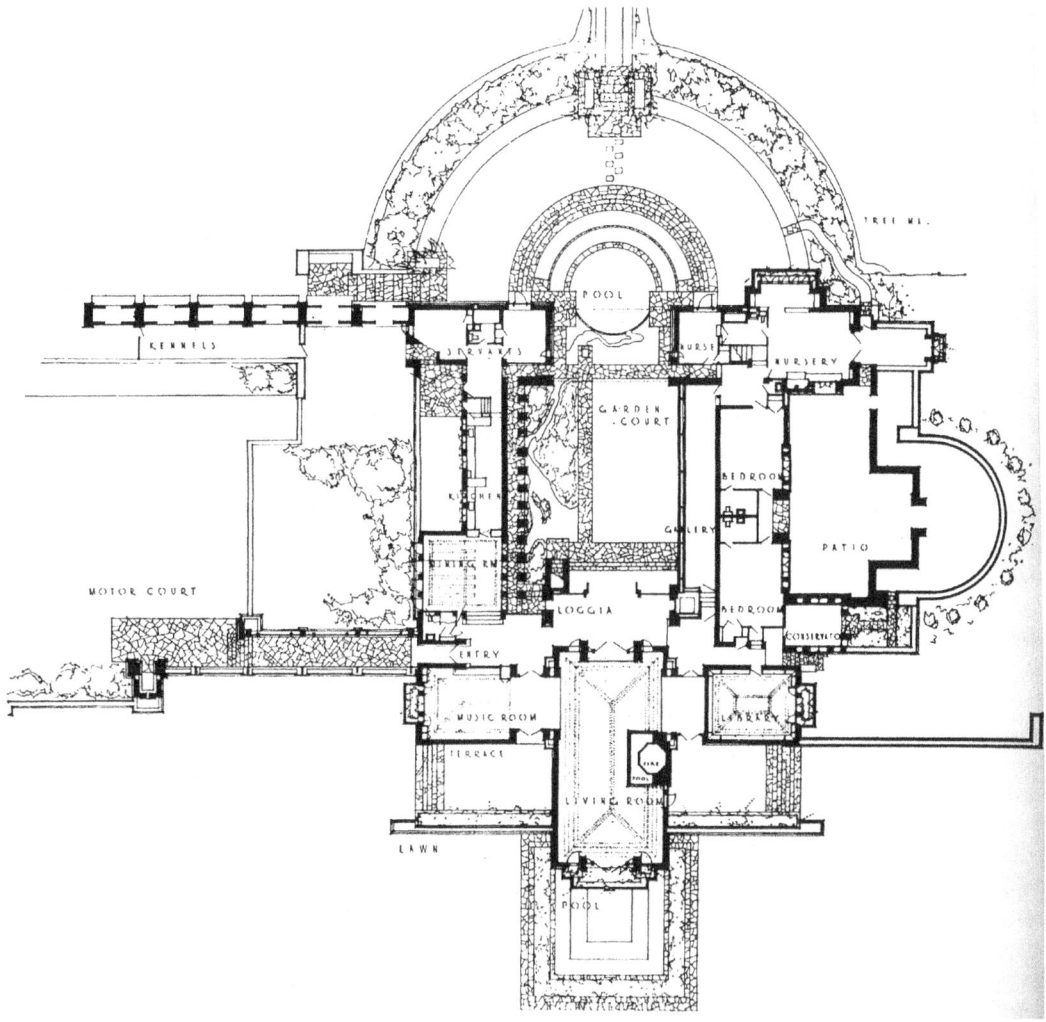

Fig. 1.2 Barnsdall house in Hollywood, California, 1919-1922

IMPORTANT EVENTS

The sequence of critical events to be studied in the following chapters, are outlined chronologically.[9]

ca.1903
- U.S. publications about concrete blocks and the machinery to make them begin earnestly, many published in Chicago.

1903-1906
- United States Shoe Machinery Company, Beverly, Massachusetts, "Daylight Factory," Ernest L. Ransome, engineer, poured concrete post, beam and slab construction.

1905
- FLW and wife Catherine and Mr and Mrs Winslow engaged in a three month tour of Japan.

1907-1911
- In San Diego architect Irving J. Gill created a series of classic modern buildings, some built of concrete, notably houses for Allen (1907), Price (1908-1909), Miltmore (1911), and Barker (1911-12); also a Boys' Dormitory for the Childrens' Home Association (1908), Scripps Institution (1908), and Bella Vista Terrace houses (1910).

1908
- The patented Turner flat-slab concrete construction system that employed "mushroom" column capitals.
- New York architect Grosvenor Atterbury perfected a system of precast concrete panels for walls, floors and roof construction and created a residential aesthetic devoid of historical referents, initially housing in Sewarn, New Jersey, to 1910.

1910
- FLW in Europe, arrived September 1909, visited Germany, France and Italy, returned to Chicago early 1911; his work published in Berlin by Wasmuth.
- (Mexican revolution began.)
- Washington, D.C. architect Milton Dana Morrill perfected a system of reusable steel forms for concrete *in situ* walls and created an aesthetic devoid of historial referents, notably for Cement City suburb in Nanticoke, Pennsylvania, to 1913.

1912
- JLW (John Lloyd Wright) and LW (Frank Lloyd Wright, Jr.) began residing in San Diego, JLW worked for architect Harrison Albright to 1913, LW with Gill to 1914.
- Chicago architects WBG (Walter Burley Griffin) and MMG (Marion [Lucy] Mahony Griffin) won an international city planning competition for what was to become Canberra, Australia's capital city.

1913
- January, FLW travelled to Japan to obtain an agreement in principle to design the Imperial Hotel and continued to study and collect Japanese art; returned in May and visited JLW, LW and Gill in San Diego.
- August to November, WBG travelled to Australia to discuss designs and plans for Canberra, he and MMG decided to reside permanently in Australia when he accepted the position of Director of Design and Construction for the new city.
- FLW's designs for Midway Gardens began, construction completed 1914.

1914
- February to March, WBG and MMG travelled to U.S. and Europe to select jurors for an international competition to design a new Australian parliament house; because of the European war the competition was postponed, later cancelled by the federal government.
- In May WBG and MMG began permanent residence in Melbourne.
- May to September, Australians George and Florence Taylor and Ernest Stowe travelled to the U.S. and July-August visited JLW, FLW and Harry Robinson, all in FLW's office, as well as architects Louis Sullivan, Francis Barry Byrne, Dwight Perkins (MMG's cousin), William Gray Purcell in Minneapolis, among others, and George Griffin, WBG's father.
- August: a crazed servant at FLW's home kills five people including FLW's lover Mamah Cheney and her children.
- Design for Walter Dodge house, West Hollywood, by Gill, completed 1916.
- (U.S. government preemptively landed troops at Veracruz, 300 Mexicans and 19 Americans killed; the U.S. Army withdrew after a few months of ineffective occupation.)

1915
- FLW visits Gill (with whom he worked in Adler & Sullivan's Chicago office ca.1890-1893); JLW, LW and FLW attended the Panama-Pacific International Exposition, San Diego.
- Schindler travelled to New Mexico, Arizona and Southern California; on return to Chicago designed the Martin house, a project for Taos, New Mexico.
- Design and construction of German Warehouse in Richland Center, Wisconsin, began using the Barton system of post and slab reinforcement, but FLW withdrew services in 1918 and German abandoned construction in 1920.
- Ellen B. Scripps house, La Jolla, by Gill.

1916
- F.C. Bogk house, Milwaukee, Wisconsin, by FLW.
- Summer: FLW officially appointed architect for Imperial Hotel.
- December, FLW began a series of extended stays in Tokyo to complete design and detailing for—and supervise construction of—the Emperor's hotel.
- WBG and MMG began design of a typical one room cottage, and design of their "knitlock" construction system.
- (Pancho Villa's raid on Columbus, New Mexico, provoked U.S. government to again send a punitive expedition into Mexico.)

1917
- WBG began procedures to patent a concrete block system he called knitlock.
- JLW employed in FLW's Chicago office (1913-1917) and Tokyo office (1917-1918); "fired" by father he left Tokyo mid-1918.

- Natco Company's "tex-tile" terra cotta blocks available, by 1919 called "textile" tiles or "textile" blocks.

1918
- February, Austrian architect Rudolph Schindler began in FLW's Chicago office where other employees were William E. Smith and from Japan Arato Endo and Goichi Fujikura.
- 30 October, FLW departed for Yokohama; then briefly visited Beijing.

1919
- FLW opened an LA (Los Angeles) office and began preliminary designs for Aline Barnsdall's "Hollyhock" house and the related Olive Hill campus in LA.
- Horatio West Court, Santa Monica, California, by Gill.
- April/May, the Taylors' envoy, Stowe, visited LW and Schindler in FLW's Chicago office, Stowe also visited Mullgardt and others in LA.
- September, FLW arrived in Seattle from Japan, travelled to LA; final scheme for Hollyhock approved by Barnsdall.
- late Autumn: construction of the Imperial Hotel began.
- 16 December, FLW departed for Yokohama, his third trip to Japan

1920
- FLW in LA from July until he departed for Japan on 16 December.
- November, Schindler transfers to FLW's LA office, LW also engaged part time, Gill a casual visitor.
- (The beginning of relative stability in Mexico, travel and research revived.)

1921
- May, FLW returned from Japan.
- June, FLW in LA, 30 July departed for Japan and remained until boarding a ship in Yokohama on 22 July 1922.
- Schindler's own house at Kings Road, Los Angeles, completed 1923.
- November, construction of Barnsdall house completed.
- November, Schindler started an independent architectural practice but continued part time for a few months with LW in FLW's LA office.

1922
- August, FLW returned to U.S. and continued LA office with LW; Frank and Catherine Wright divorced in November.
- LW designed "knit-block" for the Henry Bollman house, Hollywood, construction drawings dated 1 December 1922, construction completed early 1923.
- G.P. Lowes house, Eagle Rock, California, FLW, design drawings by Schindler.
- 1 August, FLW arrived in Seattle not to return to Japan, and did not travel to LA.

1923
- February, FLW's mother died.
- FLW and LW adapted the Griffin interlocking concrete block system and used it ineffectively on the Millard house. They then designed edge-grooved tiles and called them "textile blocks".
- February, FLW resided in LA and attempted to invigorate his practice, considered a formal collaboration with LW; employed in LA were Kameki Tsuchuira and his wife, and William E. Smith; LW was paid by Barnsdall, Freemans, Storer and Ennises.
- El Pueblo Ribera court, La Jolla, California, by Schindler, completed 1924.
- March, final design approved for Millard house, Pasadena, construction completed March 1924.
- July, FLW began design of Olive Hill kindergarten.
- September: Imperial Hotel officially opens and during opening ceremony survives the disastrous Kanto earthquake; FLW writes "Why the Skyscraper?", an essay later published as a pamphlet entitled *Experimenting with Human Lives* that contained the first but inexact public mention of his concrete block construction system.
- September, FLW closed his LA office and returned permanently to Spring Green, thereafter to occasionally visit Southern California until the late 1950s.
- November, FLW married Miriam Noel; on 24th construction stopped on Olive Hill kindergarten.
- Into 1925 —> Designs submitted by FLW to Edward L. Doheny for a resort ranch in Beverly Hills, not a commission but a speculative proposal by FLW and rejected by Doheny; FLW designed a house for Dorothy M. Foster in Buffalo, New York; FLW promoted his personal speculative but unsuccessful project (including building designs) for a resort beside Lake Tahoe; FLW designed a Phi Gamma Delta fraternity house in Madison; FLW began design of a new Barnsdall house for Beverly Hills; he designed facilities for the Nakoma Country Club near Madison, and an unrealized A.M. Johnson "compound" at Death Valley. At one stage of design each of these projects were to employ his textile blocks and tiles, none were realized.

1924
February, designs approved for Freeman house in LA, construction completed March 1925; designs approved by Ennises but soon they rejected LW's supervision of construction and executed some non-Wrightian changes, construction was completed August 1925.
- April, architect Louis Sullivan died in Chicago.
- October, Austrian architect Richard Neutra began at FLW's Spring Green home office.
- FLW began an "affair" with Olgivanna Hinzenberg; estranged from second wife Miriam.

- December, WBG and MMG arrived in LA, then proceeded to Chicago.

1925
- February, WBG and MMG in LA.
- Second major fire at Wright's Spring Green home called Taliesin.
  February, Neutra left Spring Green for LA where he continued to work occasionally in the LW/FLW office; met the Griffins either in Chicago (December 1924 to January 1925) or in LA when they were enroute back to Australia; Neutra began organizing research notes that included textile tiles and knitlock to be published in Germany in 1926 and 1927; he began design of the Lovell house in 1927.

And so forth.

PRECAST CONCRETE

Glue: the word evolved from ancient Greek to mean anything sticky. "Cement" is from the Latin to indicate a sticky substance that after it hardens acts as an adhesive, like glue. Acting as a binder, cement was and is the principle binding ingredient of concrete. That is how the Romans understood the word and so it is today. Egyptians used a cementitious material 4000 years ago but not for structures. Herod of Biblical fame built a large breakwater of enormous precast blocks composed of hydraulic concrete at the town of Caesarea on the coast of Judaea. As well it is reported that Romans used precast concrete elements to build culverts, aqueducts and tunnels. One common formula contained a volcanic ash found near Pozzouli beside the bay of Naples and below Vesuvius that they called *pozzuolana*. Pliny reported a masonry mortar composition of one part lime to four parts volcanic sand and ash mixed with small stones and water. Concrete was valuable only as a structurally compressive material. When its slurry dried it would bind rubble infills (mainly random stone or broken and full sized brick) to form bearing or non-bearing walls, pillars and tall massive arches and vaults such as the impressive great aqueduct of circa 10AD that still bisects today's Segovia, Spain.[10] Concrete was a material visually too coarse for the finish of wall, arch and vault surfaces, so fine marble or slim long Roman bricks (face or edge) were applied to wall surfaces. Often Roman bricks were the formwork that helped contain the rubble and concrete infill. The stigma of concrete's unsightly or unrefined coarseness persisted well into the nineteenth century.

After the collapse of Rome the use of concrete (of one formula or another) continued in Byzantium and thereafter in the Muslim world of Southwest Asia and northern Africa. Europeans, however, were gripped by Christian medievalism and a downturn in economic enterprise. As one consequence concrete was forgotten—or at least its potential not advanced—until the 1800s, 1300 years later. During the 1700s a few Englishmen had worked on developing a more useful natural hydraulic cement from lime. Then in 1822 James Frost prepared an artificial hydraulic lime not unlike that of Frenchman Louis Vicat's of 1812. It was commonly known as Frost's cement.[11] In 1818 the American engineer Canvass White found limestone deposits in upper New York that, with little processing, could form hydraulic cement that would harden under water. It was used safely as mortar for the stone work for the Erie Canal that ran from Lake Erie, past Buffalo to the Hudson River at Albany. It was completed in 1825.[12]

Those investigations and patents were precursors to the English bricklayer and mason Joseph Aspdin's first Portland cement patent in 1822 "for making artificial stone". Production of what he called Portland cement began at Leeds in 1824. There were, therefore, two types of cement available: one of naturally occurring materials and one artificially made by a concoction of synthetic and natural ingredients. Imports of Aspdin's cement began in 1868 and controlled the American market beyond 1871 when David O. Saylor at Lehigh Valley, Pennsylvania, obtained a patent and started a Portland cement plant, the first in the U.S. In 1880 a barrel of dry American Portland cement cost $3, but only $1 in 1900.[13]

In 1867 the London architectural firm of Shaw and Lockington proposed a set of houses made of reinforced concrete. Designs by Englishman W.H. Lascelles for cast "cement slabs" and cast concrete bricks were published in the 1870s. Some American industrial plants were built of reinforced concrete by the engineer Ernest L. Ransome beginning in the 1890s. His large United Shoe Machinery Plant in Beverly, Massachusetts, of 1903 featured precast beams set into U-slots at precast post heads. And during 1902 and 1903 the first reinforced concrete skyscraper office building was built, the Ingalls (or Transit) Building in Cincinnati, Ohio, by architects Elzmer and Anderson in association with the Ferro-Concrete Construction Company.[14]

During the 1850s Frenchmen Jean-Louis Lambot produced boat hulls, Francois Coignet filler joist floors, Joseph Monier (a gardener) flower pots and garden post. Each of these men used cast concrete reinforced with metal wire. Perhaps the first patented reinforced concrete two storey flooring system supported by beams (reported as fireproof construction) was by the English inventor William B. Wilkinson in 1859. One of his cottages of the 1850s survived until demolished in 1954.

Today Portland cement is made by heating a mixture of limestone, clay, slag and gypsum to 1500°C (2732°F) in large rotation kilns that produce a clinker. When the clinker is crushed it becomes the familiar grey powder, the grey of Portland stone found near Leeds. During the heating process, for every 100 tons (90.7 tonnes) of limestone (or calcium carbonate), forty-four tons (39.6 tonnes) of carbon dioxide are driven off to help produce fifty-six tons (50.8 tonnes) of quicklime (or calcium oxide), the major binding constituent of cement. As a filler, an aggregate is added to the mix depending on its use. The filler can be almost anything that, after being saturated with water, does not shrink or absorb the water and is salt free: coarse salt-free sand, gravel (metal), rocks, recycled concrete chips, aluminum flakes, and so on. And the water must be potable and also salt free. Today, when people refer to their cement footpath or street they in fact mean concrete. If only cement and sand are mixed without an aggregate it is often called "cement".

The neatest working definition is that by Henry Saylor in his little *Dictionary of Architecture*: "con`crete, a compound of cement, large and small aggregates, and water, deposited in temporary forms [formwork] while in a fluid state. When set it attains hardness and strength not unlike stone".[15] Polymers can be added to increase flexibility for, as examples, diving boards or a ship's hull. During the drying or curing process a great deal of heat is released over the first thirty hours. Water under plastic sheets help keep the drying concrete cool and avoid overheating and eventual cracking. Thereafter it slowly releases heat as curing continues. Interestingly, it is reported that concrete continues to harden for thirty or more years and during that period about one fifth of the carbon dioxide that resulted from the original production of cement is reabsorbed. Yet it is clear that the negative environmental impact during the process of making cement and then pouring and drying concrete mix is of staggering proportions when compared to other structural systems.

Once a slurry of cement infused with additives was put into a mold to harden as concrete, business and creative minds searched for practical applications and possibilities of shape. The potential to exploit molded or cast concrete for a variety of uses, including masonry construction, teased the imagination of engineers, constructors, architects and other designers. It is interesting to note that T.J. Lowry developed and produced a "Patent Mold for Building Blocks, 1850".[16] The concrete blocks he illustrated were almost exactly the same as Atterbury's foundation walls or today's two chambered hollow-core blocks.

Initially blocks "were made by shoveling a dry mixture of concrete into a mold and tamping it as the formwork was filled". Wright's textile blocks were hand tamped with a mallet and wood block through the 1930s. However, as historian Handlin has pointed out:

> *By the beginning of the [the twentieth] century automatic tampers replaced hand labor. Other concrete block were made by a related process in which the mold was first completely filled and then the concrete mechanically compressed . . . an improvement over tamping because it ensured a block with a more uniform density. . . .*
>
> *When concrete was fluid, the aggregates . . . became thoroughly coated, a better crystallization took place, and a stronger block produced. By 1905 manufacturers of wet blocks were beginning to outstrip those who used tamping and compression.*[17]

When Wright's blocks were used for the Arizona Biltmore Hotel in 1927-1928 a wetter-than-usual mixture was still hand tamped.

One of the earliest uses in the United States of concrete blocks above ground, so to speak, was for the George A. Ward house, New Brighton, Staten Island, at some point in 1837. A chunky building decorated on its exterior walls with English Gothic revival things molded in wet stucco, the heretofore unrecognized lumpy structure "was built wholly of cement blocks which were manufactured in square boxes and placed on top of each other" (the blocks, not the boxes), then covered with a plaster of cement or lime stucco.[18] For many years such blocks were called "cast stone" or, early in the twentieth century, "liquid stone". Apparently Horace Greeley ("go west young man"), editor of the New York *Tribune*, built a three storey barn of cast concrete at Chappaqua, New York, in 1852. In 1868 George A. Frear of Chicago patented a cast or "artificial" stone and in that year the H.B. Horton house on Calumet Avenue was completed with Frear Stone.[19]

Metal reinforcing bars or rods were introduced into the slurry during the late nineteenth century, an essential technological advancement to future applications. Steel reinforcing has the same coefficient of expansion as concrete and provides tensile strength and thereby a resistance to unwanted bending and twisting or to collapse by shear. The term "ferro-concrete" was often mentioned and, as we shall learn, in the 1920s there was a fleeting moment when it was thought that it would foster a unique architectural style.[20] Engineering studies of reinforcing were coupled with a search for applications that would exploit concrete's capacity to fireproof skeletal iron and steel structures. Potentialities *seemed* limitless.

Perhaps the first structure in the United States to be built entirely of reinforced concrete using Portland cement imported from England was by another Ward, William E. of Port Chester in New York, high above Long Island Sound. From 1873 to 1877 Ward, a mechanical engineer who conducted considerable research, supervised the construction of his own house made of "béton combined with iron", to use his words, that is rods and I-beams. Ward probably learnt something from a book by Major General Q.A. Hillore's *A practical treatise on Coignet-Bèton and other artificial stone* of 1871. The house was designed by New York architect Robert Mook who amalgamated stylistic elements of castleated medieval, Tuscan, and an Anglicized Second Empire.[21] Historical eclecticism was the vogue. To locals it was at first dubbed "Ward's Folly" but when to their surprise it remained standing it was called Ward's Castle.

Contemporary with William Ward's expensive effort and after limited local experiments in Honolulu in the 1870s, a number of buildings were built of concrete blocks. Among them were three houses and the Honolulu Post Office—each extant today—and a store designed by English emigrant architect and builder J.G. Osborne.[22]

Construction of the two Ward houses also coincided with the release in 1867 of a book in England entitled *Suburban and rural architecture. Brick, stone, concrete and fireproof*, edited by architect E.L. Blackburne. Included were some solid Italianate/Romanesque proposals by the previously mentioned architects Shaw & Lockington: perhaps W.N. Lockington of Lower Clapton. But was the other architect the Londoner R. Norman Shaw? The designs certainly have Shavian quality and proportions and his works were admired in America. In any event all were to be built in "Portland Cement Concrete".[23]

Another influence upon Mook may have been Richard Norman Shaw's *Sketches for cottages and other buildings designed to be constructed in the patent cement slab system of W.H. Lascelles*. Ernest Newton, Shaw's chief architectural assistant for many years, independently used the system.[24] Later, at the 1878 Paris Universal Exhibition a house was "Built entirely of Concrete bricks by W.H. Lascelles". Shaw was listed as architect but a drawing of the front elevation of the town house indicates he was not the author.[25] During the nineteenth century the influence of England on North America in all manner of things was, well, huge.

With Aspdin's breakthrough in the 1820s England took the lead in concrete technology and application.[26] But events on the continent eventually succeeded that lead. One notable success began with a search for an economical method of fireproof construction. In the 1880s the prolific French builder François Hennebique developed a reliable reinforcing system of metal straps. Familiar with those buildings, another entrepreneur builder, Californian English-expatriate Ernest Ransome patented twisted square reinforcing bars that would inhibit slippage when concrete shrunk while drying. As an engineer and building contractor Ransome built many large reinforced concrete structures before the turn of the century.[27] In 1893 he built Stanford University's Museum using reinforced concrete, the walls coated with stucco and painted. He then went on to develop the modular "Ransome Unit System" of precasting concrete elements used to erect entire building sections. Wright, who used the phrase "unit system" thereafter for what was no more than a simple design and construction module, surely knew Ransome's work.

The commercial application of reinforced concrete took off and other American proprietary structural systems soon followed.[28] One was developed by the engineer C.A.P. Turner in 1910; another by the engineer Julius Kahn, brother of Detroit architect Albert Kahn.[29] Julius' company became the Truscon Corporation. Another was developed by Francis M. Barton and called the Spider Web System for reinforcement of concrete slab construction. Wright used it for the structure of the Albert D. German warehouse in Richland Center, Wisconsin, begun in 1915.[30] From 1890 to 1900 there were thirty-five U.S. patents for methods of forming or casting concrete; 125 patents from 1900 to 1908; then 800 from 1908 to 1914.[31]

One method needs recounting. The fertile, commercial, entrepreneurial mind of Thomas A. Edison turned to concrete when he established the Edison Portland Cement Company in 1899. Among other projects was one to construct and sell mass produced housing in the form of detached dwellings. In 1906 he devised methods for a "Poured Cement House." A conveyor belt took suspended buckets of wet concrete to the top of the formwork where men would tip the buckets so that a thin moved downward. Ineffective, he then pumped the slurry in by compressed air. The first trial occurred in 1909 using cast iron molds. He also produced precast cement furniture (reinforced with slim wires) that included a fancy rococo-style phonograph cabinet.[32] Edison's company succeeded but the lumpy, complex houses designed "in the style of Francis I", we're told, by architects Manning and Macneille were a constructional and commercial failure.

All the technical advances in concrete could not persuade architects to expose it off-form and untreated externally or internally. The aesthetic preference was to cover it, to face it with an acceptable surface material like stone or sometimes light weight terra cotta tiles that imitated stone or bricks. But usually a cement stucco surface was tooled to resemble stone work. Exceptions after the 1870s were for

Fig. 1.3 Unity Temple in Oak Park, Illinois, 1905-1908

those building types where social and aesthetic sensibilities were not challenged: machine shops, factories, sheds, warehouses, or whatever. Yet it was exposed externally by carefully controlled mixes, formwork and pouring for the Ponce de Leon Hotel in St. Augustine, Florida, designed by the architectural firm of Carrère and Hastings, from 1885-1887. It now houses Flagler College.[33] Joseph Siry found that between 1905 and 1910 a number of Wright's professional colleagues in Chicago were designing buildings with concrete left exposed off-form; included were warehouses and an occasional park pavilion.[34]

When French architect Anatole de Baudot designed the church S. Jean de Montmartre, Paris (1897-1905), he used a system where compressive members were made of pierced brick (i.e. terra cotta) reinforced with threaded steel wires packed with concrete. The upper skeletal elements of post and their interlaced arches were concrete made with fine sand reinforced with wire and exposed on the exterior and interior.[35] Therefore, and contrary to suggestions, there is no comparison with either Griffin's or Wright's concrete block systems. The result was a Gothic edifice befitting the structural rationalism expressed by French architect E.-E. Viollet-le-Duc, Baudot's mentor.

The year the fragile-appearing S. Jean's was complete, Wright began designing the Universal Church in Oak Park, Illinois, to be called Unity Temple (see Fig. 1.3 above). He began designing in 1905 and the church was opened in 1909. Initially he intended to employ thick brick bearing walls that would support concrete beams and roof slabs. But the walls were finally (and with the design otherwise unaltered!) built of reinforced concrete poured *in situ* together with ornamental elements of precast concrete. Internal surfaces of the walls were coated with cement plaster. On the exterior walls was an exposed stone aggregate mixture troweled on to the surface that gave an appearance of untreated off-form concrete. Perhaps Wright knew of Edwin Lutyens' application around 1900 of exposed aggregates on elements of a few buildings and even an off-the-ground water tank's exterior where multicolored aggregates were exposed. In any event exposing aggregates by a wash of acid to cut away cement, or by sand blasting or by wire brushing was growing in popularity especially with Arts and Crafts people and followers of the bungalow craze.

To return to Unity Temple, Siry's book presented two valuable illustrations of drawings that had been placed on a single sheet of paper. Both are detailed elevations of a single exterior colonnette capital. On the left is a drawing of, among other angular forms, a line of leaves literal in form, life-like. The right drawing is of the leaves "conventionalized," as Wright would have said, abstracted as sets of rectangles. Siry then illustrated drawings of the molds and formwork used when casting the colonnettes and then photographs of them finished in precast concrete.[36] Wright said of the casting process that the Temple . . .

> was entirely cast in wooden boxes, ornamentation and all. The ornament was formed in the mass by taking blocks of wood of various shapes and sizes, combining them with strips of wood, and, where wanted, tacking them in position to the inside faces of the boxes. The ornament partakes therefore of the nature of the whole, belongs to it. So the block and box is characteristic of the forms of this temple.[37]

The wood formwork was therefore typical of the industry.[38] Yet the penultimate design, as noted, was for brick boxes.

With advances in structural engineering, in materials science, and in refinements to metal reinforcing and aggregates, Auguste Perret was able to follow Wright's Temple with the gothically slim and coarsely textured Nortre-Dame du Raincy in Paris of 1921-1922. Described by engineer and architecture historian Reyner Banham as "a confusing monument, but of considerable importance", it "confirmed Perret's standing as master of reinforced concrete in the eyes of a generation that was convinced that new material would revolutionize architecture".[39] It too was constructed of reinforced concrete pour *in situ*: floors, posts, columns, walls and roofs. Impressions of the removed wood formwork ("shuttering" in England) on surfaces were boldly left exposed, unpainted, throughout the interior and exterior. Non-structural wall elements as infill between columns on exterior walls were made of concrete precast in small *square* molds, most with embedded glass, and stacked on edge like masonry: as Wright would do. Thus, Perret used concrete as a structural frame and also for precast panel infills, glazed or not.

During the first two decades of the twentieth century concrete construction in North America gained popular acceptance for utilitarian and engineering works such as bridges, warehouses and dams, and a growing appreciation by a minority of architects and clients. The American Concrete Institute and the Concrete Block Machine Manufacturers Association were formed in 1905. Technical and promotional magazines kept people abreast of advancements and were good publicity. New York's *Cement Age* began in 1904 and in 1915 merged with Detroit's *Concrete* and Cleveland's *Concrete Engineering* to become *Concrete*. Those out of Chicago included *Concrete and Engineering* that also recognized the importance of presenting European ideas. It began July 1896 to continue beyond 1970. Started in 1907 *Cement World* was subsumed in 1917 by *Engineering World*. Also in about 1907 *Cement Era* began by promoting cement machinery. Of course the Chicago architectural and construction periodicals *Inland Architect*, *American Contractor*, and *News Record* regularly included concrete applications. The Chicago-based Portland Cement Association began publishing its *Proceedings* in 1916 and introduced the magazine *Concrete Builder* in 1918. During the previous decade the association began production of promotional and technical pamphlets on all aspects of the plastic material. These included subjects such as "Competitive designs for concrete houses" in 1907 and "Concrete Ships" in 1917.

Chicago architect Marion Lucy Mahony, who at the moment worked part time for Wright and later married architect Walter Burley Griffin, entered a design for a suburban residence to be constructed of concrete blocks in a Universal Portland Cement Company sponsored competition of 1908.[40] The same company published a book about cement houses in 1910 that illustrated Wright's house for Edmund D. Brigham in Glencoe, Illinois, of 1908 to ca. 1910.[41] It was, the company said, constructed of "monolithic" reinforced concrete floors and walls.

After the turn into the twentieth century how-to-do-it picture pamphlets and books appeared quite frequently. Most notable for our purposes, although aesthetics was seldom considered, were *Souvenir of the Shaw Block Company . . . made of cement concrete . . .* published in Chicago in 1904 by R.R. Donnelley & Sons; Henry Wittekind's *Concrete Block Houses* of 1905 in Cleveland, Ohio; Paul Wilkes' pamphlet of about 1905 on *How to Manufacture Concrete Hollow Blocks*; Spencer B. Newberry's *Hollow Concrete Block Building Construction* was produced by the Chicago based *Cement and Engineering News* in about 1905; H.H. Rice and others wrote *Manufacture of Concrete Blocks and Their Use in Building Construction* and published by the Engineering and news in 1906; the Sears, Roebuck company began a series of catalogs illustrating their blocks and block machines; then *Concrete-Block Manufacture; Processes and Machines* by H.H. Rice was published in New York by J. Wiley & Sons in 1908; Charles Palliser's *Practical Concrete-Block Making . . .* appeared in 1908 by the Industrial Publication Company of New York; William A. Radford with his own and very popular publication house "Architectural Co." produced *Cement Houses and How to Build Them . . . perspective views and floor plans of concrete block . . .* that was released around 1909; the Atlas Portland Cement Association published *Concrete Houses and Cottages* in two volumes in 1909 and redprinted them in one volume in 2010.

In 1908 *House Beautiful* magazine proclaimed "The Age of Concrete" was in full swing. In an editorial they optimistically exclaimed that, "you [the reader] may build a factory or a chicken house of it, a railroad or a piece of statuary, a canal, except for the water, and a boat to sail on it; a church . . . and images of the saints". After the Erie Canal's natural cement lining proved successful the material was then employed in a number of canals up to 1850s when concrete was used thereafter, notably for the Panama Canal. Of things concrete the magazine found a "garden table recently made is a thing of beauty;" there were "Egyptian sphinxes which look as real as any sphinx you ever saw;" and there were cupids and "weeping Magdalens" and an "American eagle all ready to scream".[42]

Wright designed two product exhibits, not pavilions, for the Universal company, a subsidiary of U.S. Steel Corporation. Little is known of the first of 1901 for the Pan-American Exposition in Buffalo, New York. Design must have been completed late 1900 or early 1901 to allow for construction and installation for the 1 May opening.[43] The other was a small exhibit for the New York Cement Show held at Madison Square Gardens in December 1910. There Wright used concrete, cement stucco, and precast concrete tiles that were applied to a low walls wide enough for seating and embracing a display table. His design was meant to illustrate "possibilities in the decoration of concrete surfaces by means of inlaid colored glass and tile". The prescribed floor space was 24 feet wide by 12 feet deep and backed to a wall. Wood molds and tiles were used similarly by Wright in 1913-1914 on the Midway Gardens in Chicago.[44]

Cast concrete details and photographs of Midway Gardens were illustrated in *Cement Era* and in the Chicago Portland Cement Company publication *Beauty and Utility in Concrete*, both of 1914. Earlier, similar information about Wright's concrete "monolith" Unity Temple, at the time still under construction, was published in *Cement World* in 1907 and *Concrete and Engineering* in 1907 and 1909.[45] A National Conference on Concrete House Construction met in Chicago in 1920. That same year the Concrete Publishing Co. of Chicago produced *Concrete Houses and How They Were Built* edited by Harvey Whipple. It included a useful section on block walls. That was just before Frank Lloyd Wright Junior began to detail his own concrete "knitblocks" that would intrigue his father.

ATTERBURY

The search in England and America for inexpensive housing enevitably settled on the new material of reinforced concrete. It could provide both structure and enclosure in one process. But the *in situ* method was too costly in labor, materials and time. The innovative English civil engineer John Alexander Brodie, Liverpool city engineer, responded to the problem by inventing a system that promised an efficient industrialized process that began in a factory where testing and quality control were possible, and ended in a quick method of construction. His idea was to prefabricate large panels by casting concrete in molds and then lifting them into position. The panels were made of cement, chips of clinker brick as aggregate, steel reinforcing, five inches thick and of varying width and length (some eleven feet square) whether floor, roof or wall. Each panel weighed tons.

Brodie's panels were apparently first applied on an unknown house in Liverpool. The second use was for a house in the Cheap Cottage Exposition begun July 1905 at a site beside Letchworth Garden City. Possibly the first British ideal homes exhibition it attracted considerable attention. While some warehouses and tram stables were constructed, the most notable building to use Brodie's system was "Concrete Tenement Dwelling, Eldon Street", Liverpool in 1906; Brodie the designer and builder. The method of construction was difficult to say the least. Panels for the tenement were cast in Liverpool, then transported by rail to the site. This may have been typical. But before construction could proceed it was necessary to build a heavy timber frame 15 feet higher than the eventual building and surrounding it. This held a traveling crane that lifted panels off rail cars and placed them to form the building. The skeletal timber frame dominated the site like a technological monster. And Brodie's system was noticed for it was soon independently developed in eastern Europe, Scandanavia and America. But not in Britain where there was strong trade-union opposition.

Perhaps during one of many travels to Europe, around 1905 or 1906 New York architect Grosvenor Atterbury witnessed Brodie's buildings under construction, or he received advice or read an article. In any event he would have understood that the biggest problem toward further development of the panel system was the huge timber frame and its crane. Surely it was possible to reduce the size of panels and thereby allow for a smallish crane, one to serve the entire site. To this end Atterbury obtained a series of patents related to cast concrete molds and systems of construction, the first application in 1908 and granted 1910, the last in 1917. The resulting panels were hollow and of limited size to reduce weight and ease of placement.

Initially with his own money Atterbury began factory production of reinforced, tongue and groove "hollow sectional" concrete floor, wall, and roof panels cast in metal molds. The panels were trucked to a building site and there assembled. Initially the principle wall panels were 3"-9" (1.43M) wide, eight feet tall (2.44M) and nine inches (0.23M) deep, hollow with thin one-and-a-half inch walls. Each weighed up to about three tons (even though cinder was used as an aggregate to lighten the composition) and a single tall electric crane lifted panels off trucks and into place. With his "Standardized Sectional System" Atterbury was "enthusiastically" trying to "bring rentals within the means of the poorer labouring classes", he said, by reducing construction and, once inhabited, maintenance costs. The building material therefore was selected for durability, economy and fireproofing. He and Brodie were the first to conduct practical experiments in mass housing using industrial techniques.[46]

Using his System Atterbury built efficient "workman's" cottages at Sewaren, New Jersey, during 1908 and 1909, see Fig. 1.4. Chastely modern (an aesthetic he was not to repeat) and nicely proportioned, they contained small rooms for kitchen, dining, living at ground level; above were bedrooms and bath. Foundation walls were constructed of concrete blocks two by three feet and ten inches thick (609x914x254mm) made at a factory on site. Electrical conduits were cast into the panels and holes left for plumbing. There was no ornament, just acid brushed exterior surfaces that exposed aggregates with subtle color and texture. Interior surfaces were smooth and painted, exteriors water proofed.[47]

Fig. 1.4 Two prototype houses at Sewaren, New Jersey, 1908-1909, Grosvenor Atterbury, Architect

Atterbury had received financial assistance from the Carnegie Steel Corporation (for reinforcing bars, crane structure and large panel molds), and beginning in 1907 (and into 1918) from the Russell Sage Foundation and its Homes Company. The Foundation's premier housing project was Forest Hills Gardens, a posh and restricted Long Island suburb begun in 1912.[48] The Forest Hills street and allotment plan was designed by landscape architect Frederick Law Olmsted, Jr., with Atterbury as architect. Similar to most of his other designs, the Forest Hills buildings were a romanticized cross of English and German Arts and Crafts flavors of the recent past. Near the "town square" a hotel and adjacent commercial buildings were pseudo-Bavarian of late medieval character in multihued brick. They emulated the town centers of proposed or built garden suburbs (that peculiar out-growth of Garden Cities ideas) as found in England, Holland and Germany. In 1912 and 1913 ten row houses forming Slocum Crescent were built using Atterbury's Section System. Even though two and three storeys and with party walls their construction was the same as at Sewaren but aesthetically different.

Precast concrete blocks and other standardized building components were manufactured at the Long Island site.[49] On the exterior surfaces of the panels aggregates were embedded, composed of crushed roof tiles, bright pebbles, bits of quartz and mica. When the surface was washed with acid the aggregates were exposed and "lent color and sparkle" to what would have been an otherwise dry, dull-grey finish. Construction time for the first row house at Forest Hills was six weeks, but the last few nine days.[50] As can be imagined, set up costs on site for

Brodie and Atterbury systems was very expensive. As a result their anticipated production potential was not realized. After a subsidy to build the row houses ceased and on completion of his coordinating and design role at Forest Hills, the Standardized Housing Corporation was formed with Atterbury and the landscape architect and city planner John Nolen as consultants.[51] But the system was too costly, the company soon dissolved.

MORRILL

Atterbury's large panels represented one of only two methods for the application of concrete, i.e. by precasting in a mold. The other was of course pouring a viscose mixture into custom made forms, usually of wood, at a building site; an *in situ* process. One of the more successful inventions for this method was designed and patented by Washington, D.C. architect Milton Dana Morrill in 1909 and termed a "sectional mold for concrete walls".[52] A series of flanged steel plates 18 or 24 inches (457 or 609mm) square (or longer if joined) were set up where a wall was to be constructed or "moulded" to use his term. Five or six plates were lined up either side of the wall under construction. Beginning at one end reinforcing and concrete were then applied. By the time workers reached the other end of the plates, concrete at the beginning would be firm and its plates relocated at the other end and construction continued. The portable steel plates were used repeatedly and rented by other builders. Morrill claimed that each working day a wall around a house could rise four feet. Morrill had studied architectural design in the prestigious Eugene Duquesne atelier attached to the *École des beaux arts* in Paris. After formal education and before 1909 he had been a designing architect with government so he easily received the odd federal commission.[53]

What intrigues us here is a housing project, amusingly called Concrete City, designed and constructed by Morrill beside the town of Nanticoke, Pennsylvania.[54] Twenty double houses, their term, were placed on a 37 acre (14.9 hectare) tract and on four sides of a generous rectangular park, playground and garden area (see Fig. 1.5). Access to the units was by diagonal roads at the park's corners to a circumferential road. Construction ran from September 1911 into 1913 as housing for workers of the Truesdale Colliery of the Delaware, Lackawanna and Western Railroad's coal division. They believed this construction technique would prove economical but the cost of $2500 per family unit was unfavorably compared to $2100 for a typical brick house. After 1921 control of the coal field, colliery and housing were taken over by the mighty Glen Alden Coal Company who promptly abandoned it all in 1924. Today the site and buildings remain a targeted ghost town and dump. The site plan was unique for its day and for today. Equally so, the units' exteriors were remarkable as severe modernist designs, the interiors plain.

As can be imagined Morrill's architecture theory was straight forward and practical, attempting to shake free of the nineteenth-century notion of style, something he learned at the *École* but now rejected, at least for his concrete buildings.

> *After considerable study I have come to believe that concrete is the material presenting the greatest possibilities from the standpoints of sanitation,*

Fig. 1.5 "Concrete City" housing Nanticoke, Pennsylvania, 1913, Milton Dana Morrill, Architect

permanence, and economy.... Economy compelled me to put out of mind all architectural development and go back to first principles and to primitive habitations....[55]

What architectural style are we developing today?... What new type and style will be developed through the use of reinforced concrete? There must be a concrete style.... [P]ossibly the best concrete buildings have been designed by engineers, as they have followed the simplest and most logical shapes, and have not been hampered by architectural precedent.

[T]he simplest and most natural forms for concrete [can] still be beautiful in proportion, line and color....

I have found a box house was by far the most economic form ... enclosing a space as this form requires the least wall area....

[S]traight lines are ideal for this work, and ... make the simplest and most attractive buildings.[56]

And so forth.

Morrill also built houses or workers' housing in Brentwood, Maryland; Virginia Heights (plain boxes and "Craftsman" style), Arlington, Chevy Chase Village, and Middleburg (ten houses he named "concrete city") in Virginia; Overlook Colony in Clayment, Delaware (75 units); New Canaan, Connecticut; Washington, D.C.; High Lake, Chicago, Illinois; Gary, Indiana, in collaboration with architect D.F. Creighton;[57] Youngstown, Ohio; New York City; and where else? Two and three storey row houses had covered porches or not and were rather industrial looking or English Arts and Crafts or American Craftsman. Houses were simple, boxey and chaste, without ornament, with covered porch or not, with trellis or not, pitched roof or flat with garden and trellised or not. The reductive aesthetic found at Nantacoke's

Concrete City was not always followed. A customer's preference for a style was now and then persuasive and pitched roofs common: but not so the few commercial, industrial and governmental commissions.

What also intrigues is that over the years Morrill returned to familiar haunts in France, best known in 1921. Surely he knew of Tony Garnier's housing schemes of 1914 and as published in 1918 that were so similar to the earlier Concrete City buildings. In 1921 the French government was still grappling with post-war housing and village redevelopment in the "Devastated Regions". That year they invited Morrill to France to erect a demonstration house in Tourcoing near the Belgium border. Apparently it proved economically viable and Morrill began promoting his system and no doubt handed out pictures of completed buildings. There are suggestions that his system was used thereafter in France but a pursuit of the possibility is outside the scope of these essays.[58] So too its influence on French or European modernist architects.

Since their architectural designs were published simultaneously on America's west coast and elsewhere, it is easy to believe that California architect Irving Gill, soon to be discussed, knew of Morrill's system and designs and Morrill knew of Gill's work in off-white stucco and concrete. A comparison of Morrill's buildings of 1911-16 to Gill's of 1908 (note particularly his Price house of 1908-09), tease the historian.[59]

The two decades before August 1914 saw bewildering activity in methods of concrete application. For another example: a complete precast, reinforced concrete constructional and fabrication process also called a "Unit System" was patented in 1912. It supplied precast building parts from foundation to roof and all between: lintels, jambs, floors, beams, wall panels, window frames, and so on. More complete than Ransome's unit system, it was designed by the engineer John E. Conzelman whose Unit Construction Company went on to erect a number of buildings across North America. Another constructional method developed by the military at about the same time and still used today was fabrication at the construction site of reinforced concrete wall panels and roof panels. A steel framework and mold rested horizontally on the ground with reinforcing in place and into which was poured a concrete slurry. When cured the concrete walls were tilted up into position while smaller roof panels, although not always used, were lifted into place.[60]

The George Ward house on Staten Island was somewhat typical of future one-off buildings that produced concrete blocks at a construction site. The practical mass production of blocks off site, ready to use, was realized when Chicagoan Harmon S. Palmer developed a hand-operated machine around 1890. Refinements to it were made while he constructed six houses during 1897.[61] Once mold and compression apparatuses became readily available at reduced prices, of course the public rapidly accepted "hollow"—that is not solid—concrete blocks. The term "hollow" referred to two or three empty or hollow chambers in each block. By 1905 about 100 companies made block machines.[62] "[P]ractically unknown in 1900," said one author in 1906, "at the present moment more than a thousand [1000] companies and individuals are engaged in their manufacture in the United States".[63] The easy availability of low cost American-produced Portland cement was a critical factor to commercial exploitation.

Wright's center of professional operations was a rapidly developing commercial and industrial metropolitan and regional giant. It was therefore a center for concrete production including blocks. A representative of the Universal Portland Cement Company, formed in Chicago in 1896, wrote in 1908 that the Chicago area was

> *the distributing point of more barrels of Portland cement annually than any other city in the world. . . .*
>
> *The output of cement in Chicago and its contiguous territory is approximately 7,000,000 barrels per year, or about one third of the output of the whole country.*[64]

Chicago's Sears, Roebuck and Company produced a catalog in 1907 that illustrated seventeen different hollow concrete blocks produced by Triumph or Wizard machines. Block faces ranged from plain to ornamented as an imitation (in relief) of cobblestone, to a variety of ashlar stone, and so on. Atterbury also invented a concrete mixture of cement asbestos fibre, sand and crushed cinders into which, when cured, nails could be driven: he called it "Nailcrete".[65] In 1917 F. J. Straub patented a block whose aggregate contained lightweight cinder and it too could be nailed. That year he made 25,000 blocks; by 1926 he was making 70,000,000 annually.[66]

Wishing to call itself the Athens of the South, from 1895 to 1897 the city of Nashville built a full sized replica of the Parthenon and placed in the center of the Tennessee Centennial Exposition campus. Not intended to be permanent the Tennessee temple was constructed of plastered wood and stuccoed brick veneer. Yet, thirty years later the city decided the temple needed a complete rebuild and selected local architect Russell Hart to supervise. Reconstruction was in reinforced concrete and finished in 1925, but only the exterior; the interior ready for occupation in 1931. Many of the building's ornamental elements were made of precast concrete and all of the exterior was exposed concrete of a buff color "not unlike" the original marble (they thought) as it had weathered over the centuries. In fact, when construction was complete in 432 BC the Greeks had painted their Parthenon in bright, joyful colors, inside and out.[67]

In spite of technical achievements and aesthetic challenges, during the 1920s and 1930s officials of the American Institute of Architects believed to visually expose concrete and concrete blocks, structural or non-structural, was aesthetically unpleasant, coarse, too obviously fake stone, and urged architects to avoid them undressed. Wright once described precast concrete (including blocks) as "one of the insensate brute materials that is used to imitate others".[68] Regardless, he used it often, poured *in situ* or precast, naked or clothed, from 1894 through the 1950s.[69]

NOTES TO TEXT

Portions of this text were first presented in a paper on invitation to the Melbourne venue of the University of Illinois (Urbana)/University of Melbourne/Power House

Museum (Sydney) international symposium on "The legacy of the Griffins: America, Australia, India," in September 1998. Another portion was prepared as a paper entitled "Griffins' knitlock, Natco's tex-tiles, Wright's textile-tiles" and presented twice, first to the 1998 Melbourne conference of the Society of Architectural Historians, Australia and New Zealand. A rather teasing paper, it was later published as "Dear Marion" in Johnson (1998). The second occasion was to a symposium, "European modernism and Frank Lloyd Wright," held in Adelaide, October 1998, organized by the University of South Australia in conjunction with the local chapter of the Royal Australian Institute of Architects. Considerable research has been accomplished in the fourteen years since those presentations and certainly since relevant material was brought to light in Johnson (1977). Aspects of the later papers were partly advanced in Johnson (2005), within chapters 1 through 4, and in Johnson (2012).

The most complete and useful bibliographic tool is Langmead (2003), with over 4000 entries that include information on reprints: it replaces a long out-of-date but still useful Sweeney (1978). And there is Steinerag.com on the web.

Illustrations, many also in color, and some construction details of all of Wright's principle buildings discussed herein can be found in Pfeiffer (14-24). The Arizona Biltmore Hotel and San Marcos-in-the-Desert (with detailed drawings of textile blocks of 1927-1928) can be found in Pfeiffer (24-36).

Drawings by and for Frank Lloyd Wright are copyright © 2013, The Frank Lloyd Wright Foundation, Taliesin West, Scottsdale, AZ.

    one mile = 1610 metres or 1.61 km
    one metre = 3'-3 3/8"
    one foot = 305mm or 30.5cm
    one inch = 25.4mm

NOTES

1   Donald Hoffman, *FLW's Fallingwater. The House and Its History* (New York: Dover, 1978, 2nd edition 1993); Hoffman (1992); idem, *FLW's Dana House* (Mineola: Dover, 1996).

2   Kathryn Smith, "FLW, Hollyhock House and Olive Hill 1914-1924" *JSAH*, 31(March 1979), pp. 15-33; her first published survey. Barnsdall commissioned Wright to design 41 buildings: 16 stores and apartments, 21 houses, one apartment building, one pavilion, one kindergarten, one movie house. She commissioned Schindler for another dozen.

3   Alofsin (1993); Nute (1993); Smith (1992); Siry (1996) and (2008). For a study of the educational, religious and philosophical influences on FLW prior to 1917 see Johnson, forthcoming.

4   Mary Shelly, introduction to the 1831 edition of her *Frankenstein or, the modern Prometheus* (London, 1818, 1831, London: Claremont 1994), p. 8.

5   On Wright's relationship with the FBI see Johnson (2005).

6   Smith (1992), title page verso.

7   Rebori (1927), p. 453.

8   Scully (1988), pp. xviii-xxi, and based on Scully (1960).

9   The time line was constructed using many sources but see Smith (1985), pp. 296-310; Nute (1993), pp. 184-186; Alofsin (1993), p. 288.

10   For an outline with specific reference to concrete see William MacDonald, "Some implications of later Roman construction", *JSAH*, 4(Winter 1958), pp. 2-3.

11   Cf. the valuable Frank Newby, editor, *Early Reinforced Concrete* (Aldershot: Ashgate/Variorum, 2001). For contemporary historical detail up to 1838, some research referred to part two of Charles W. Pasley, himself a cement manufacturer, *Observations on Limes* (London: 1828).

12   Harley J. McKee. "Canvass White and Natural Cement, 1818-1834", *JSAH*, 20(December 1961), pp. 194-197; Robert W. Lesley, *History of the Portland Cement in the United States* (Chicago: International Trade Press, 1924), pp. 13-14, a comprehensive review; [Charles E. Peterson], "Poured concrete building, 1835", *JSAH*, 11(May 1952), pp. 23-24, the article also reported that the August 1835 issue of *Mechanics' Magazine and Register of Inventions and Improvements* described a "little structure of poured cement just erected in New York City", and that in 1844 "poured concrete buildings" were built in Wisconsin.

13   Wermiel (2000), p. 167.

14   Carl W. Condit, "The first reinforced-concrete skyscraper: the Ingalls Building . . .", *Technology and Culture*, 9(January 1968),pp. 1-33, reprint Newby (2001), pp. 255-292.

15   See also the entry "reinforced concrete" and compare with "precast concrete" in *Encyclopedia*; and for the science see Sidney Mindness and J. Francis Young, *concrete* (Englewood Cliffs: Prentice-Hall, 1981); Theodore H.M. Prudon, "Simulating Stone 1860-1940: Artificial Marble, Artificial Stone, and Cast Stone", *APT Bulletin*, 2(no 3/4, 1989), pp. 9-91; and for a brief review Ada Louis Huxtable, "Concrete technology in U.S.A., Historical Survey", *PArch* (October 1960), pp. 143.150; for a better review see Collins (1959).

16   Handlin (1979), patent illustrated on p. 280.

17   Handlin (1979), pp. 279-280.

18   This is a discovery, so to speak, of a concrete block house not previously cited in historical texts, Atlas (1909), vol. 1, illustration p. 9; but see brief verbal description in Collins (1959), p. 56.

19   [Charles E. Peterson], "Frear Artificial Stone, Patented 1868," *JSAH*, 13(March 1954), pp. 27-28.

20   Onderdonk (1928), throughout.

21   Slaton (2001), p. 16, 165; Wermiel (2000), p. 165; Jandl (1991), pp. 55-65; Collins (1959), p. 57; Ellen W. Kramer and Aly A. Raafata, "The Ward House: a Pioneer Structure of Reinforced Concrete," *JSAH*, 20(March 1961), pp. 34-37, reprint Newby (2001), pp. 251-254. The house required 4000 barrels of Portland cement imported from England. Mook was best known for his cast iron Anglo-Italianate commercial buildings of the 1870s and 1880s in New York City's So-Ho district. Until recently the mansion housed the Museum of Cartoon Art.

22   [Charles E. Peterson], "Concrete Blocks, Honolulu, 1870s," *JSAH*, 11 (October 1952), p. 27. Osborne had a thriving practice in Hawaii.

23   *Suburban and Rural Architecture. Brick, Stone, Concrete and Fireproof*, E.L. Blackburne, editor, (London: James Hagger, 1867); idem, editor, *Suburban and Rural Architecture, English and Foreign* (London: James Hagger, 1869), with color lithographed plates.

24 R[ichard]. Norman Shaw, *Sketches for Cottages and Other Buildings Designed to be Constructed in the Patent Cement Slab System of W.H. Lascelles* ([London]: W.H. Lascelles, 1 May 1878). Newton received the RIBA Gold Medal in 1918.

25 <www.auctiva.com/hostedimages>, accessed 2 April 2007.

26 R.J.M. Sutherland, "Pioneer British Contributions to Structural Iron and Concrete: 1770-1855", *Building Early America*, Charles E. Peterson, editor, (Radnor: Chilton Book Co., 1976), pp. 96-118.

27 Reyner Banham, "Ransome at Bayonne", *JSAH*, 42(December 1983), pp. 383-387; Banham (1986); Collins (1959), pp. 61-72.

28 Slaton (2001), pp. 138-193.

29 On the history of concrete construction techniques related to architecture generally see Slaton (2001); Wermiel (2000); Collins (1959); Banham (1986); Ford (1990); Onderdonk (1928); Herbert (1986); Newby (2001); and Carl W. Condit, *American Building Material and Techniques* (Chicago: University of Chicago Press, 1968), p. 169ff. As developed by Wright see Patterson (1994); Frampton (1995), Chapter 4; and for the period 1890-1910, Siry (1996), pp. 108-111, 294, n91-96, n98.

On Kahn see Frederico Bucci, *Albert Kahn. Architect to Ford* (Milan: CitaStudi, 1991; Princeton Architectural Press, 1993), n31-37; Grant Hildebrand, "New Factory for the Geo. N. Pierce Company, Buffalo, New York, 1906", *JSAH*, 29(March 1970), pp. 51-55, place of the Pierce Arrow; George Nelson, *Industrial Architecture of Albert Kahn, Inc.* (New York, 1939). Contemporary British developments are *outlined* in A.C. Davis, *A Hundred Years of Portland Cement 1824-1924* (London, 1924); outlines on those French and German see Collins (1959); Frampton (1995), Chapters 3 and 5.

For a pictorial survey of the variety of uses and of building types for concrete in history see William Hall, editor, *Concrete* (London: Phaidon, 2005).

30 Eaton (2006), pp. 264-269.

31 Onderdonk (1928), p. 32.

32 A less ornate yet historically derived designed followed almost immediately and a few were built in New Jersey. Frank Lewis Dyer, et al, *Edison his Life and Inventions* (New York: Harper, 1929), vol. 2, chapter 20; *Scientific American*, "The Edison Concrete House," (August 1909); Collins (1959), pp. 90-91, plate 27A; Ronald W. Clark, *Edison the Man Who Made the Future* (London: MacDonald and Jane's, 1977), pp. 184-187, 199; Matthew Josephson, *Edison* (New York: McGraw-Hill, 1959), pp. 423-425 and illustrations.

33 Carl W. Condit, "The pioneer concrete buildings of St. Augustine", *PArch* (September 1971), pp. 128-133. On the proliferation of domestic buildings that used Portland cement for reinforced concrete or as stucco on lath or masonry see Atlas Portland Cement Co., *Concrete Country Residences* (New York, 1906), 4th edition 1909.

34 Siry (1996), pp. 109-110.

35 John Jacobus in Macmillan; Collins (1959), pp. 113-117, plates 31-33. On St Jean de Montmartre see G.J. Edgell, "The remarkable structures of Paul Cottancin", *The Structural Engineer* (London), 63A(July 1985), pp. 201-207, reprint Newby (2001), pp. 169-186.

36 Siry (1996), pp. 150-151.

37 FLW, "In the cause of architecture. IV: fabrication and imagination", *ARecord*, 62(October 1927), p. 319, with thanks to Siry (1991), p. 152; Eaton (2006), pp. 255-259.

38 Siry (1996), figures 97 and 98.

39 Reyner Banham, *Theory and Design in the First Machine Age* (London: Architectural Press, 1960), pp. 41-42; Collins (1959), pp. 202-219; see also Kenneth Frampton, *Modern Architecture a Critical History*, 2nd edition (London: Thames & Hudson, 1985), chapter 11.

40 Universal Portland Cement Co., *Plans for Concrete Houses* (Chicago: 1909), figures 50-53; "Architectural Prize Competition," *American Contractor* (Chicago), 29(November 1908), p. 75, from Siry (1996), p. 294,n91.

41 Bingham house was first published in Universal Portland Cement Co., *Representative Cement Houses* (Chicago, 1910); Storrer (1993), pp. 186-187. Wright's design of a "monolith" bank project in 1901 was for a brick building stuccoed and then that was slightly altered for concrete.

42 As quoted in Chusid (1992), p. 66.

43 The exposition closed on 2 November and is best remembered as the site of President William McKinley's assassination on 6 September. Almost immediately on closure all structures were demolished except the New York State Pavilion.

44 Alofsin (1993), pp. 167-169, 342,n35. The quotation: Universal Portland Cement Co, *Monthly Bulletin*, No. 79(December 1910), p. 3, as cited in Siry (1996), p. 306,n27. Wright Archives drawing 1004.01 is a plan and elevation, 02, 04, 05 are details of wood molds. First illustrated in Wright (1911), n.p.

45 Kruty (1998), pp. 33-34; Mary Woolever, "Prairie School works in the Ryerson and Burnham Libraries at the Art Institute of Chicago," *Museum Studies* (the Institute), 21(2, 1995), p. 144, cf. p. 108; Siry (1996), p. 143, 154-155.

46 Herbert (1986), p. 16; Onderdonk (1928), figures 67-69.

47 Opposite Staten Island and just north of Perth Amboy, Sewaren is only 259 hectares and 2700 or so people and part of Woodbridge township. The houses were built on land owned by Sewaren Improvement Co., a real estate company; snippets of information from many online sites, October 2012.

48 Letter Sage Foundation Homes Co. to W.E.B. Du Bois, 2 November 1912, rejecting on grounds of race Du Bois' attempt to purchase a building lot; <library.umass.edu/view/full/mum310-b6007-il36> accessed September 2012.

49 On Atterbury's concept for Forest Hills see his "Forest Hills Gardens, Long Island, an example of collective planning, development, and control," *Brickbuilder* (New York), 21(November 1912), pp. 317-318. On his concrete construction systems see idem, "Studies in economic construction," *Cement Age*, 2(December 1910), pp. 315-325 (well-illustrated), revised as, idem, "The Economic Production of Workingmen's Houses," in "Buildings: their uses and the spaces about them," *Regional Survey of New York and its Environs*, vol. 6, 1931; Frederick Squires [and G. Atterbury], "Houses at Forest Hills Gardens—Pre-cast Hollow Concrete...", *Concrete Cement Age*, 6(June 1915), pp. 3-8, 55-56 (also well illustrated, the issue was devoted to concrete and concrete block houses); Standardized Housing Corporation, *The Manufacture of Standardized Houses: A New Industry* (New York; 1917); Peter Pennoyer and Anne Walker, *The Architecture of Grosvenor Atterbury* (New York/London: W.W Norton, 2009); Klaus (2002), pp. 55f, 77-78, 185,n39,n.42; Handlin (1979), pp. 287-288; W.F. Anderson "Forest Hills Gardens—Building Construction," *Brickbuilder*, 21(November 1912), pp. 319-320; Jandl (1991), pp. 24-26.

50 Klaus (2002), p. 78.

51 Standardized Housing Corporation, *The Manufacture of Standardized Houses: A New Industry* (New York: the corporation, 1917), illustrates how the Sewarn houses were

built and the lifting equipment; Forest Hills buildings are not illustrated. See also Grosvenor Atterbury, *The Economic Production of Workingmen's Homes ... obtained from researches and demonstrations, 1904-1925* (New York: Russell Sage Foundation, 1930).

52 Morrill's patent submissions were dated 1909, 1910 and 1911 and in Canada 1913. He may have set up the business in Brooklyn, New York, as Read and Morrill, Inc.

53 Milton Dana Morrill, *The Morrill moulded concrete houses* (New York: the author, [1919]); idem, *Inexpensive homes of reinforced concrete* (Philadelphaia: the author, 1910); reprint Morrill (1910), pp. 86-88; reprint National Association of Cement Users, *Proceedings,* 6(1910), pp. 468-493 as a paper read to the Association convention Chicago; reprint American Concrete Institute, *Journal of Proceedings,* 6(February 1910), pp. 468-490; reprint *Western Architect,* 16(October 1910), pp. 103-106; reprint *The American City,* 3-4(June-July, 1910-1911), pp. 177-180, a paper read before the Second National Conference on City Plannning.

See also W.A. Du Pay, "The Healthiest House in the World", *American Homes and Gardens,* 6(July 1909), pp. 269-270; Morrill's comments in Whipple (1920) where Irving Gill and Morrill are in proximity. On a survey of industrial workers' housing see Atlas Portland Cement Co., *Industrial Houses of Concrete & Stucco* (Chicago/New York: the company, 1918), the title page has a color perspective drawing of the Nanticoke housing.

In 1911 Duquesne, then architect for the French government, accepted the appointment as professor of architectural design at Harvard University in Boston.

54 "Nanticoke" was derived from Nantego, name of the tidewater Indian people who were forced to move north out of Maryland. See Robert A. Janosov, "Concrete City, Garden Village of The Anthracite Region: *Pennsylvania Heritage,* 23(Summer 1997), pp. 32-40; and <www.itsveryeasytorememeber.com/.../history_of_concrete_city> accessed September 2010. The Cement City suburb in Donora, Pennsylvania, built 1916-1917 by Lambie Concrete House Corporation for the American Steel aand Wire Company, had two storey New England style bungalows. And there is Cement City village in Michigan.

55 Milton Dana Morrill, "Inexpensive homes of reinforced concrete", *Proceedings of the second national conference on City Planning and the Problems of Congestion",* (Boston/Cambridge, Massachusetts: University Presss, 1910), p. 83. Atterbury also spoke at the convention.

56 Morrill (1910), pp. 87-88.

57 Charles Vinz, "The [American] Sheet and Tin Plate Company ... Gary, Indiana", *Proximity Magazine,* no. 3 (Winter 2008-2009), n.p.

58 "American concrete houses at Tourcoing", *Housing Betterment, 1921-1922,* vol. 22, [U.S.] National Housing Association, pp. 22-23.

59 For a contemporary history that includes concrete *in situ* see Winthrop A. Hamlin, *Low-cost Cottage Construction in America, A Study Based on the Housing Collection in the Harvard Social Museum* (Cambridge, Massachusetts: Harvard University, 1917); James Ford, *The Housing Problem* (Cambridge, Massachusetts: Harvard University, 1911); Whipple (1920). For a brief but useful city planning and housing context for the Atterbury and Morrill systems see Margaret Crawford, *Building the Workingman's Paradise: The Design of American Company Towns* (London/New York: Verso, 1995), p. 31ff.

There were other steel formwork systems but it seems they began operation after Morrill's: Metaform 1912, Schub 1912, Lambie in 1915, Van Guilder 1916 or earlier, and others.

60   J.L. Peterson, "History and Development of Precast Concrete in the United States," *Journal of the American Concrete Institute*, 25(February 1954), pp. 483-484. Tilt-up construction remains an often used process.

61   Walton D. Stowell, "The 'Miracle' and the 'Wizard', Preliminary Notes on Concrete Block Machines", APT *Bulletin of the Association for Preservation Technology*, 5(no. 2, 1973), pp. 67-70; Ann Gillespie, "Early Development of the 'Artistic' Concrete Block: The Case of the Boyd Brothers", ibid, 11(no. 2, 1979), pp. 30-52, block faces are illustrated; Theodore H.M. Prudon "Simulating Stone, 1860-1914," ibid, 21(no. 3, 1989), pp. 79-91.

62   With thanks to Professor Miles Lewis. Many products to follow were apparently in breach of Palmer's patent.

63   S.B. Newberry, "Hollow Concrete Block Building Construction in the United States," *Concrete and Constructional Engineering*, 1(May 1906), p. 118.

64   James P. Beck, " 'King Portland' and the Great Central Market," *Cement Era*, 6(April 1908), p. 94, as quoted in Siry (1996), p. 109.

65   Klaus (2002), p. 77.

66   On Straub see Pamela H. Simpson, *Cheap, Quick, Easy. Imitative Architectural Materials 1870-1930* (Knoxville: University of Tennessee Press, 1999), chapter 1; and idem, "Cheap, Quick and Easy: The Early History of Rockfaced Concrete Block," in Thomas Carter and Bernard L. Herman, editors., *Perspectives in Vernacular Architecture, III* (Columbia: University of Missouri Press, 1989), pp. 108-118.

67   John J. Earley, *The concrete of the architect and sculptor* (Chicago: Portland Cement Assoc., 1926), pp. 7-10; Earley and the sculptor Lorado Taft were vigorous supporters and users of concrete for sculpture and architectural ornament. There are many texts on coloration of Greek buildings and sculpture but see Matthew Gurewitsch "True colors", *Smithsonian* (Washington), 39(July 2008), pp. 66-72.

68   Wright (1928), p. 102.

69   Thomson (1999), for good visual material. In about 1947 Wright designed a "Cement Block Plant for W.E. Gifford", although Gifford denied knowledge of the design or receiving drawings, see Mary Jane Hamilton, "The Builders' Company concrete block plant", Sprague (1990), pp. 173-177.

It is uncertain if Wright intended to use natural adobe for newspapeman Lloyd Burlingham's projected house of 1941-1943, to have been located in desert highlands outside El Paso, Texas, Wright called the proposal "pottery house"; see plan, section and colored perspective drawing Pfeiffer (2011); Riley (1994), pp. 268-269. [David A. Hanks?], *Concrete masonry: The Work of FLW* (McLean, Virginia: National Concrete Masonry Association, ca.1976), was not sighted.

2

The Buildings

> Here is my theory of structure [i.e. of architecture]: A scientific arrangement of spaces and forms to functions and to site—An emphasis of features proportioned to the *gradated* importance in function—Color and ornament to be decided and arranged and varied by strictly organic laws—having a distinct reason for each decision—The entire and immediate banishment of all makeshift and make believe.
>
> — **Horatio Greenough, 1851**

Wright's father and the Jones family of his mother believed the greatest of men were Abraham Lincoln and Ralph Waldo Emerson. The poet and philosopher Emerson, his followers and the Unitarians were, of course, read and quoted and thoroughly appreciated throughout America. The sculptor Horatio Greenough and Emerson were close friends. The above epigraph was within a letter to Emerson written on December 28th of that year. Its content was quickly endorsed by the recipient in a January 1852 reply: "Well, joy, & the largest fullest unfolding to your theory!" The theory was a portent of the philosophic thrust of North American and European architectural modernism in the first half of the twentieth century; an oft repeated by-product was the reduced aphorism "form follows function". When Greenough's hypothesis was attached to nationalism that outlined the theoretical ideas Louis Sullivan uttered around 1900. Brought up on a familial feed of Emersonian and Unitarian principles and then, Frank tells us, weaned on Sullivan's critical comments, the theory became Wright's touchstone. Yet as we shall learn, it was not always upper most in practice.

FOUR LOS ANGELES HOUSES

Almost immediately on return from Japan in late 1922 Wright began experimenting with what he later called "textile blocks" or "textile tiles" or less often "textile-block-slabs" of precast concrete.[1] Initially the blocks and tiles were most effectively used as facing or structural members for walls and rectangular posts on four houses: first for former client Alice Millard, and then for new clients John Storer, the Freemans, and the Ennises. Each house was built on steep hillsides and located at the suburban perimeter of a quickly expanding Los Angeles: a city centered on undulating desert lands, high lumpy hills and seasonal arroyos; the first major city to exploit the linear movement of trolleys, trams, and automobiles. Like Barnsdall, those four clients, those patrons were an interesting lot.

Alice Covell Parsons, a Chicago school teacher, was twenty-eight when, in London, England in 1901, she married Chicagoan George Madison Millard, age fifty-four; they were together on a business trip. George collected and sold books privately and also managed McClurg's rare book department of the Chicago store. In 1905 Wright had prepared architectural plans for Millard's first house in a rather typical Wright Style of the period. It had a functionally logical floor plan, plastered interior walls, a regional appearance with stained horizontal wood board and batten siding and a shallow pitched roof. Construction was complete in 1906 on a site in suburban Highland Park north of Chicago. Perhaps because of George's health, around 1913 the Millards sought the dry climate of Pasadena, California, where they carried on the book business. After George's death in 1918, only a couple of years after formal retirement, Alice included European antique furniture and decorative arts to what had become a thriving business. She became "the single most influential antiquarian bookseller in Southern California in the 1920s and early 1930s," said star cultural historian of Los Angeles, Kevin Starr. He went on to provide readers with an insight into Millard's social acumen and aura:

> *Alice Millard became tutor to 1920s Los Angeles, a direct conduit of [British] Edwardian culture and civility to the Southland. Her Burne-Jones tresses now coifed and rinsed in surreal blue as a mark of election, Alice Millard functioned as a priestess of taste for an affluent era struggling to shake off its Babbittry.*[2]

Edward Burne-Jones was a founder of England's profit-minded Pre-Raphaelites who were active from the 1850s into the 1890s and precursors to the Arts and Crafts Movement.

Alice's tutorials were conducted from a new Wright-designed house very different to that of 1905. Late in 1922 the architect eagerly pursued his former client for a number of months, insisting on providing her a special house, "the best in [his] portfolio", he teased, to be built of his "novel system of construction", as she put it. His enticements even offered to forego the standard architects fee for share of future profits when the house was sold.[3]

> *Her house should be entirely sensible [he said retrospectively]. Should she have one of those Spanish blisters with a paper roof, with Spanish decorations pasted on it? She should not; nor should she have anything of the scene painting the [1915] San Diego Exposition favored. She should have a real home....*
>
> *Instead of a fire-trap for her precious book collections and antiques, my client would have a house pretty well fireproof.*[4]

Wright began designing in January 1923 and the house was constructed using Wright's special concrete blocks from March 1923 to April 1924 on a cup of land surrounded by hills ragged with eucalyptus.[5] See Figs 2.1 and 2.2. (Floor plans were

Fig. 2.1 Alice Millard house in Los Angeles, California, 1923-1924

based on Wright's earlier design of 1916 for Frederick Bogk's house in Milwaukee, Wisconsin.[6]) When a separated addition to Alice's house was constructed during 1925 and 1926 to a design by Wright's son Lloyd (father had managed to alienate Alice), using a few of the same blocks here and there, it functioned as a display gallery with a large storage area, and guest apartment. She called it her "Little Museum of the Book". In it she mounted a number of exhibitions well into the 1930s. She also maintained the company name of George M. Millard Rare & Fine Imported Books until her death by cancer in 1938.

Around 1931 Alice became a close friend of Estelle Doheny, a collector of first editions and wife of Los Angeles heavy weight oil magnate Edward L. Doheny, a

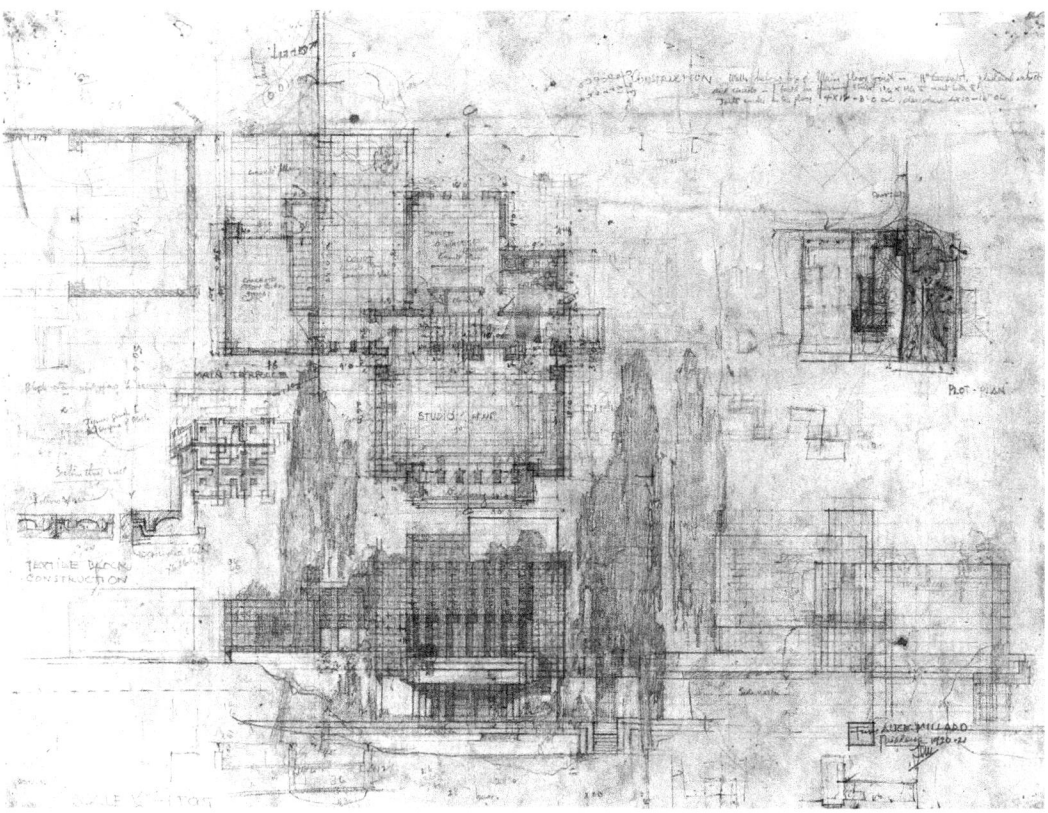

Fig. 2.2 Millard house, preliminary floor plan of entry or second level ("studio"), 1923

man wrongly linked to Wright. Edward donated money and Estelle 4000 rare books and manuscripts to a Doheny Library, the new main library at the University of Southern California in Los Angeles. She also helped finance St. John's Seminary in Rancho Colleguas Hills.[7] As we shall learn, Edward was aloof to Wright. In a memoir friend Lucille Miller observed that Alice seemed ...

> *to be all intellect, all soul, all mind. ... The ordinary, the common place, the second-rate, were not for her. Ugliness and vulgarity gave her actual pain. ... [Alice] was a Voice crying in the Wilderness —the Wilderness of the Depression; the Wilderness of provincial class-conscious Pasadena; the Wilderness of her struggle to survive in a business as a woman alone, without capital. ... But she was undaunted and undismayed. She ... expounded her Gospel of Beauty to any who would listen.*[8]

On death Alice's collection was bought by friends and given to the Huntington Library.

While Wright wrote enthusiastically and at length in his 1932 autobiography about Mrs Millard and her Pasadena house, he said nothing about the other block houses except in one instance where he linked them.

> *Months before [Millard's] La Miniatura [house] was finished improvements and changes were made in block technique. Other block houses began to grow up now on the "block system" as the "first" was being completed. The Storer House*

> was one. The Freeman House another. And then the "Little Dipper", kindergarten for Aline Barnsdall, which she destroyed, half way. And she employed my superintendent of Hollyhock House [Rudy Schindler] himself—by the way, he was ready--to turn it into a garden terrace.
>
> The Ennis house was fifth of the block-shell group. I had drawn my son Lloyd into the effort and after completing the plans and details for this latter house I entrusted it all to Lloyd to build—and went back to Taliesin [in Spring Green].

The deceptive term "block-shell" was used at least thrice in his autobiography.[9]

The next house to consider was for John Storer. He may have practiced homeopathic medicine in Chicago before moving to the Los Angeles area around 1917 but records of the American Medical Association have no listing of Storer. Perhaps it was in Chicago that he came to know Wright's architecture and professional stature, or they may have met. In any event it is rumored he turned to real estate in 1919 after failing to pass the state medical licensing examination. Little else is known about the "short and heavy set" man that Wright called "the Doctor." Storer's house, designed August 1923 and constructed of Wright's textile blocks, was ready for occupancy in October 1924. It sits on the outside of a sharpish road-curve scrunched against and embraced by the wall of a steep hill and overlooked by houses above (see Figs 2.3, 2.4 and 2.5). Interestingly, Storer put together the Superior Building Company in 1921 perhaps as part of a real estate venture. The company was indicated as the client on Wright's architectural drawings.[10]

Apparently uncomfortable in a radically different house, in 1925 Storer sold it after barely a year in residence. Thereafter the highly textured split-level house knew a number of owners. In 1931 Pauline Schindler, still separated from Rudolph, was renting the house and loved it.[11] A recent resident, Hollywood film producer Joel Silver bought the house in 1984 and gave it a needed and careful near-perfect restoration.[12] Assembled by Silver, the team leader was Wright's grandson, architect Eric Wright; restoration architect was Martin Eli Weil; then president of Los Angeles Conservancy; interior design was coordinated by Linda Marder; and concrete restoration work was by Peter Buren.[13]

Samuel Freeman, who owned a jewelry shop, and wife Harriet Press, a New York trained teacher originally from Nebraska, were married April 1921. Her sister Leah Press Lovell and husband Philip M. were witness. Harriet became a locally well-known dancer committed to the modern dance movement and "seriously pursued progressive social causes."[14] They were good friends of Aline Barnsdall and impressed by plans for her mansion they hired Wright. In February 1924 they bought a very steep building lot overlooking Holllywood. Less financially affluent than Wright's other three clients the Freemans accepted a small cave-like house designed in February 1924 with construction complete in March 1925 (see Figs 2.6 and 2.7). Encouraged by Harriet their new home, built of Wright's textile blocks, quickly became a salon of local avant-garde activity that carried on into the 1970s. As well, Hollywood people black listed by the U.S. House of Representative's Un-American Activities Committee in 1947 found intellectual stimulation and social sanctuary.

Fig. 2.3   Alice Millard house in Pasadena, California, 1923-1924

Fig. 2.4   John Storer house in Los Angeles, California, 1923-1924

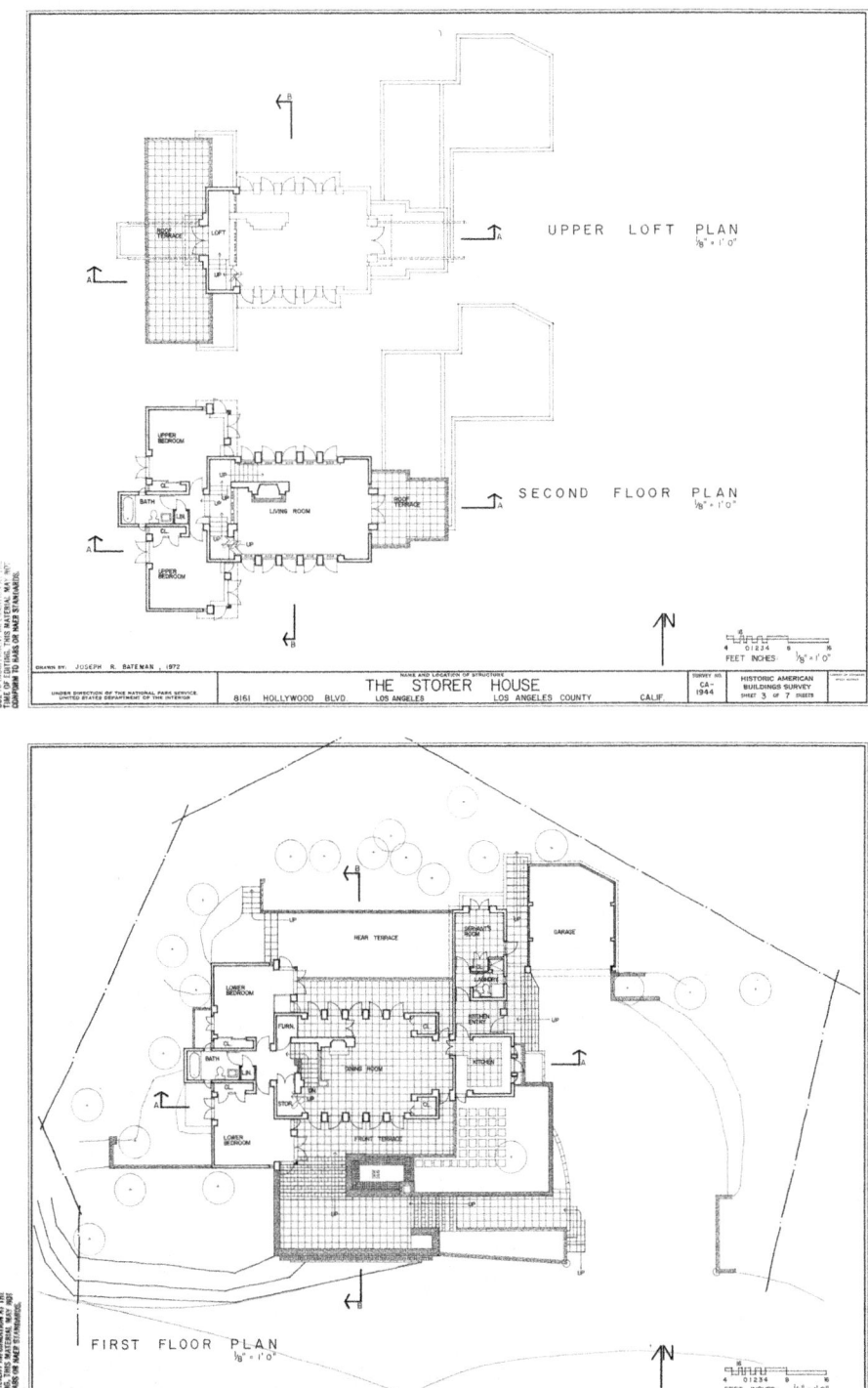

Fig. 2.5 Storer house, measured as built drawings, 1972

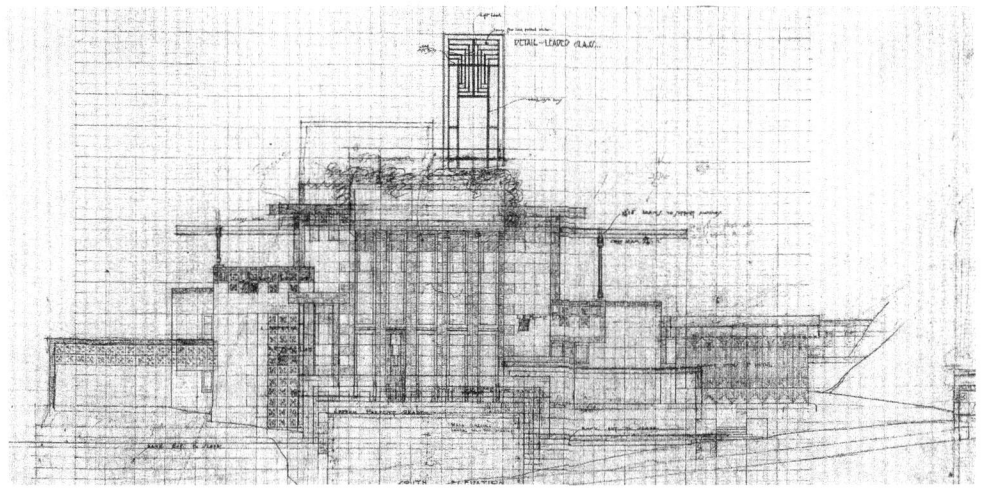

Fig. 2.6 Storer house, preliminary elevation drawing of 1923

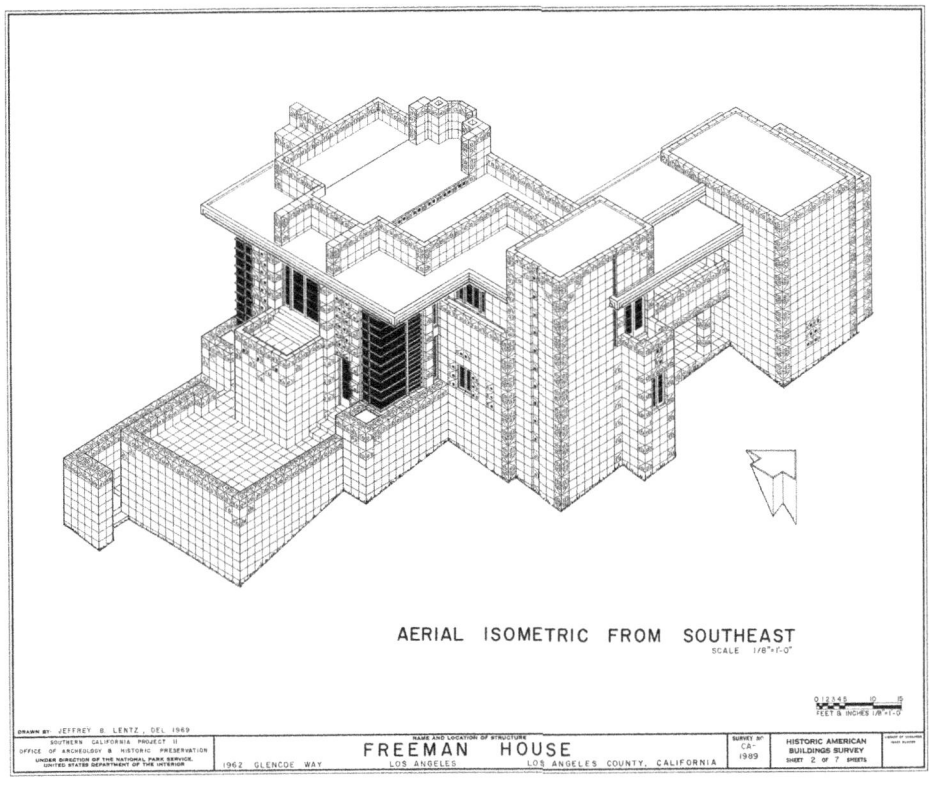

Fig. 2.7 Samuel and Harriet Freeman house on a hillside above Hollywood, California, 1924

One of the blue collar construction workers, Byron Vandergrift, who occasionally worked for Lloyd Wright as a laborer and draftsman, made some of the blocks and tiles on site. He laster recalled that: "I used to have to carry the cement and sand from where they'd dump it on the street [and] then mix it by hand.... They had a small sledge hammer... and we put a two-by-four block on the sixteen-inch-square thing and pounded it down ... [Lloyd] showed me how much stronger the blocks got if you watered them a couple of times a day for three weeks".[15] The entire house was cut into a precipitous hill of soft soil and directly overlooked Hollywood spread out below. Although separated in the late 1920s, the Freemans lived there for the rest of their lives. Well aware of the importance of their house in the history of architecture and of Los Angeles, Harriet bequeathed it to the school of architecture at the University of Southern California in 1986. After years of neglect by USC and the effects of annual earthquakes that shifted footings and foundations, cracked walls and of decayed blocks, funds from Getty Conservation Institute, Federal Emergency Management Agency (FEMA), Los Angeles Conservation, and other donors, were forthcoming during 2000 and 2001. Those funds insured a very expensive "seismic retrofit" that included deep concrete caissons beside and under parts of the house, a long deep retaining wall on the roadside, and replacement of thousands of crumbled concrete blocks.[16]

Charles and Mabel Ennis were not local but had moved from Pittsburgh, where he was a manager for the department store Joseph Home and Company, to the Los Angeles area around 1901.[17] At the time of commissioning Wright to design a house Charles operated a men's store of "absolutely correct clothes" in the bustling city center. Like so many Southern Californians the wealthy Ennises enjoyed entertaining and apparently were more than casually interested in Mexican art. Their long and massive two bedroom house designed in February 1924 was constructed of Wright's textile tiles and complete in August 1925. It sits high in the Los Feliz hills on a prominent site overlooking Griffith Park and bound on three sides by an arcing road that rises steeply south to north where cars can enter a motor court. It can be seen miles away from Barnsdall's Hollyhock as a large grey cubic mass (see Figs 2.8 and 2.9). In 1926, only two years after retirement and less than a year after the house was complete, Charles died. Mabel continued living in her special mansion for another ten years, selling in 1936.[18]

Wright Junior supervised construction of Barnsdall's mansion as well as the four houses. Nonetheless Wright Senior remained involved enough to managed to irritate and then alienate each client. The Ennises and Wright had a tempestuous relationship such that in December 1924 the Ennises refused supervision by the Wrights.[19] They assumed control at a point when construction had reached only a few feet above the entry court level and changed floor and ceiling materials and a few fixtures.[20] Storer and the Ennises had no further dealings with architects, Wright or anyone else. But when it came time for architectural assistance later on, the Freemans, who were nearly destitute after refinancing their house (and Wright had willingly assisted them in some manner) sought out Schindler and Millard hired Wright Junior. Charles died in 1929 and in 1935 Mabel sold the house. In 1968 Mr. and Mrs. August O. Brown became the eighth owners of the Ennis house.

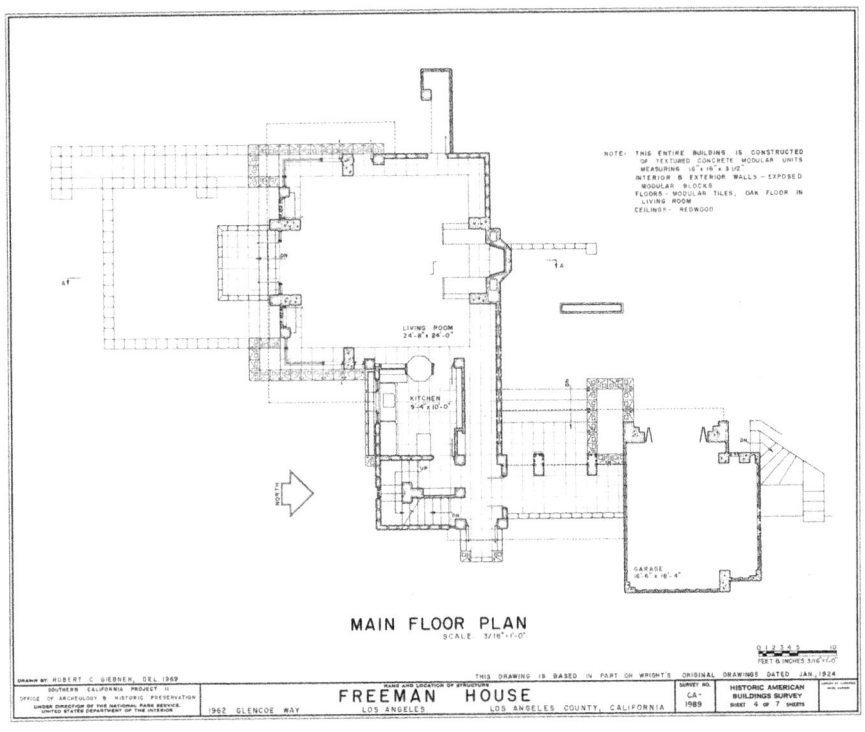

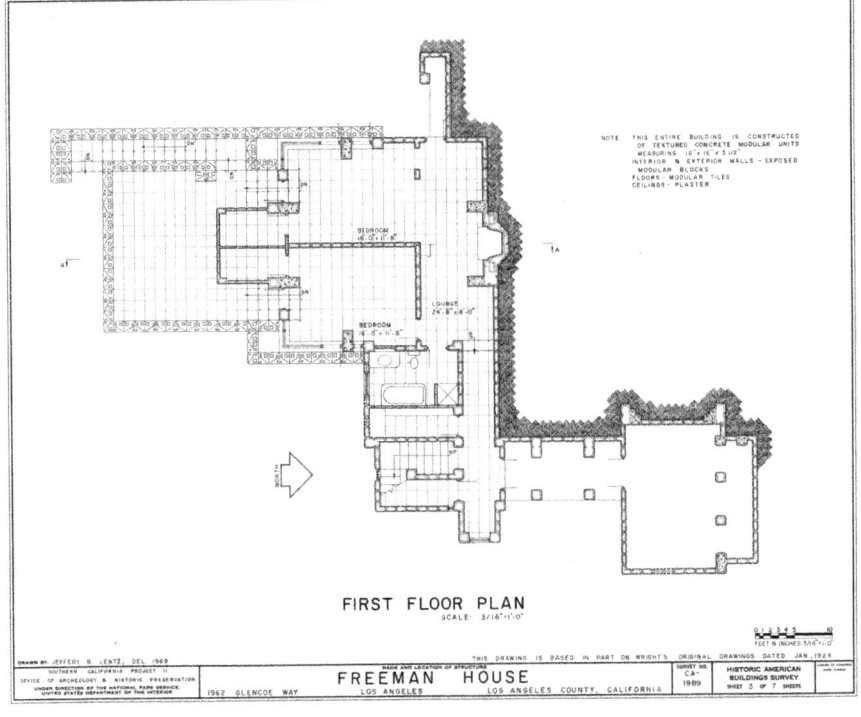

Fig. 2.8   Freeman house, as built drawings, 1969

Fig. 2.9 Freeman house in Los Angeles hills, 1924-1925

Around 1980 Mr. Brown established a non-profit corporate Trust for Preservation of Cultural Heritage "to insure maintenance of the house in perpetuity, and donated the house" to the Trust.[21] But those funds were insufficient. The Ennises house had suffered earthquakes, but that in 1994 and subsequent torrential rains were disastrous. The same donors that assisted restoration of the Freeman house came to the rescue.[22] The shoring-up price was 6.5 million. Interestingly, because of its unique appearance, since 1932 the house has been used in scenes for at least thirty-five movies and a few television commercials. Most notable were some smokey claustrophobic scenes in Ridley Scott's classic *Blade Runner* of 1982 for Warner Brothers.[23]

Harriet Freeman and her sister Mrs. Leah Press Lovell were active in Barnsdall's theater, avant-garde art and politically leftist circles. Philip Lovell was attracted to alternative medicine and acted as a health columnist for the *Los Angeles Times*. Leah was an early disciple of the Italian/American progressive educator Angelo Patri who was a follower of John Dewey's. Leah was a cofounder with Pauline Schindler of a private nursery school supported by Barnsdall and located on Olive Hill in Residence B. Being sympathetic to socialist and commensurate left-wing ideas, the Freeman's and others of Barnsdall's clique were often challenged and damned by the intolerant of Los Angeles society. For example, while the clique may have agreed with many of the ideas put by the outspoken Emma Goldman, described as a radical anarchist, she was ridiculed or hated by most Los Angeleños. Barnsdall considered her a philosophic sister and when Goldman was in Los Angeles she was accepted as a guest at Hollyhock. Barnsdall and Goldman believed that theater was a powerful vehicle to expose social injustice by inference.

Fig. 2.10   Charles and Mabel Ennis house in Los Angeles, California, 1924-25

Late in the 1920s the Lovells had a radically modern house built to designs by Rudolph Schindler (he and wife Pauline were out-spoken leftists) in concrete and glass, and by the personable Richard Neutra, his design of steel frame and glass now rightly famous. Both architects were former Wright employees. By 1927 Wright had long before left Southern California and was fiddling with his textile blocks for a hotel in Phoenix, Arizona. The odd client of the five was the conservative Storer who fit the profile of many—in fact most—of Wright's clients in previous decades who were not alternates or progressives or avant-garde-ists.

A KINDERGARTEN

Except for a small pumping house, aspects of plan, space, and application of material on the two residential buildings for Olive Hill, called A and B during the years 1919 into 1921, were architecturally ambivalent. Although principally designed and detailed by Schindler, Wright (who was in Tokyo on most of the important occasions) was involved just enough to render the buildings confusing in all aspects. Their walls were standard construction in the use of hollow clay tiles and/or wood studs and lath. Each material had a stucco finish on the outside and plaster inside. We can understand that, with the absence of Wright in Tokyo or Madison, Schindler had no choice but to make design decisions and keep Wright informed of them by mail.

However, there was a fifth textile block building contemporary with the four houses during that creative season of 1923 and 1924. Leah Lovell and Pauline Schindler had been running a kindergarten on Olive Hill with a curriculum based on the Patri method that emphasized free-association and learn-by-doing. Shortly after Wright's return from Japan and before Barnsdall departed for another trip to Europe she expressed a desire to establish a kindergarten for her daughter and for neighborhood children who would, of course, pay tuition. It was called a Community Playhouse. That title followed the lead of preschool educators who, to perhaps over simplify, believed that, aside from taught activities, various forms of play and participation were to young children as brain storming was to adults. Pauline and Leah were no doubt Barnsdall's counsellors as was the teacher she hired, Helen E. Girvin. The program Girvin and Barnsdall agreed upon fitted nicely with the progressive education agenda then gaining wide acceptance in urban America and to which Pauline and Leah were dedicated. In fact a number of Wright's clients financially and purposefully supported progressive education, notably the Coonleys and Martins. With the title "Creative Education" a one-page prospectus was issued and in part said:

> Nature would have her children be children before they are men [adults]. . . . The project of Olive Hill has this spirit as its basis—not to provide another unit striving for new fads among the myriad schools organized and uprising throughout the country, but to help the children by the wisest and most gifted guidance available to find themselves.[24]

And so forth. The curriculum was to include typical subjects plus French (in songs), music with piano and violin lessons, dancing, cooking, hygiene, nature trails, art

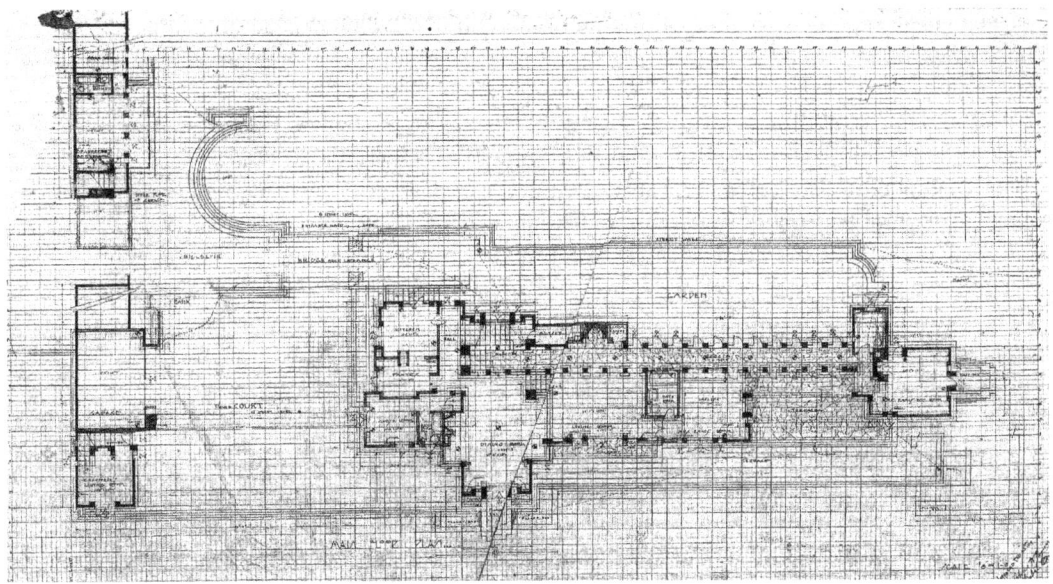

Fig. 2.11 Ennis house, 1924-1925, main floor plan

and visiting galleries, and Saturdays were devoted to civics and visits to museums. It was a six day program and similar to one Wright would have known through his aunts Hillside Home School. It was also very similar to that given to Wright by his clients of twelve years earlier, the Coonleys. Their private "playhouse" kindergarten in Riverside, Illinois, also catered to local children.[25] Barnsdall would have known of the Coonley kindergarten through Pauline if not from other sources. Before marriage Pauline worked with Jane Addams at Hull-House during 1917 and into 1919 teaching the arts. She of course knew of Wright's many buildings.

A small site on Olive Hill was surveyed in July 1923, initial preliminary designs were made during August and into September with the final design probably in October. The kindergarten was designed to fit awkwardly on a steep slope half way down the west side of Olive Hill and between two dirt roads. There was to be a single school room with a tiny one-step platform as a stage in front of, strangely, a large fireplace. (The stage was not in the final design.) Beyond the room and cut into the hill was to be a hemicycle of stepped stone and grass to act as seating to embrace a flat area of sand for an outdoor classes and play. But it would have needed a ten foot (three metre) high retaining wall on its southern perimeter. Wright named the building Little Dipper (see Fig. 2.10). Walls were to be textile blocks of six different face designs and at least thirty block variations were planned for! Although the plan parti was similar to the Millard, Freeman and Storer houses, the kindergarten appears in its structural and aesthetic complications (we can say "over designed") to be the odd one of the five built that season, or rather this one only started. For reasons still unclear—costs or demands of building codes or indecisions by client or architect or whatever—construction of the kindergarten was halted on 22 November 1923, eleven days after work started. Concrete foundations were in place as well as 226 precast concrete tiles. Left in a pile were 7282 unused tiles. The kindergarten held in Residence B ran for about three years.

Fig. 2.12
Kindergarten
on Olive Hill
in Hollywood,
California, 1923,
for Aline Barnsdall

Around this time Barnsdall had become irritated and disenchanted, as had other Los Angeles clients, with a worried, flighty and testy Wright trying desperately to reignited a flagging architectural practice. Still, he tried to save the project and reduce the pile of unused concrete blocks. He suggested a large memorial to Aline's father. But that gimmick came to naught. In July 1925 Barnsdall asked Schindler to finish the project by using the in-place foundations and installing a public garden, a "wading pool and pergola" as a place for personal contemplation.[26] The city eventually accepted Barnsdall's gift of about one half of Olive Hill land and then let the garden feature go to ruin. It was a bit of wreck when visited by this author in the 1980s. Jiggled to instability by the 1994 earthquake it was demolished.

PLAN PARTI

A building's floor plans, especially the main or entry floor, will invariably reveal a conceptual parti. In this case the parti also reveals how closely related are three of Wright's houses and the playhouse. In Fig. 2.11 a module of squares, and in some instances the principal room, are indicated by an X. We can see that the plan parti of the entry floor of each building is controlled by a dominant square room or, at Storer, a room with a module of two squares. During the prairie years 1898 to 1914 the square was Wright's modular tool—some might say an aesthetic tool—for two and three dimensional architectural and town planning and for ornamental design. In Los Angles he returned, we can say, to the square.[27] We can also see that ancillary functions are disposed off one face of the principal space with a garage attached. Roof terraces are located on some of the ancillary rooms. Functions are stacked in

Fig. 2.13 Plan parti of the Millard, Freeman, and Storer houses, and Barnsdall kindergarten

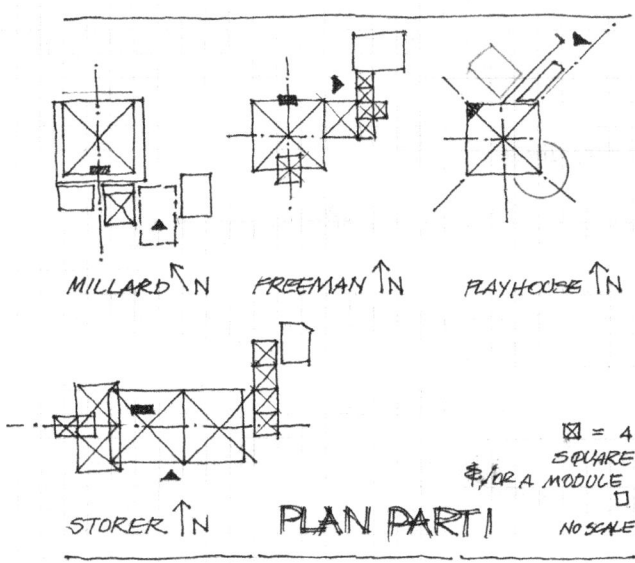

response to the steep ground for Millard and Freeman, the principal rooms and their module are one above the other: dine over living room at Millard, living/dine over two bedrooms at Freeman. Millard's bedroom is located on the third storey above ancillary spaces. Since Storer is cramped between road and hill, on the long axis of the dominant module the principal spaces are one over the other: dine below, living above.

The playhouse was to be one storey awkwardly (impractically) placed on steep land with the entrance stepping down to the play room or up to a small roof terrace. The parti does not reflect the site's disadvantage. Its principal axis is on the square's diagonal while the three houses have what might be termed a normal axis.

Millard's house faces more or less north so the entry and ancillary spaces receive some high sun, the main rooms in shade, winter and summer: Freeman's and Storer's face south. Millard's is protected by tall trees from strong winds; Freeman's and Storer's receive winds or breezes of some relief in summer. The playhouse school room was to face the bright southwestern afternoon sun that would shade the pool; the building exposed to wet or cool winds.

The Ennis house is quite different to the compact parti just discussed. It is a house in two distinct parts. Servants quarters above the garage are a two storey building separated from the house proper by a paved entry "motor court". On the opposite side of the court is the pedestrian entry to the house proper. There are two storeys about the entry with adjacent ancillary rooms and an office/study. The functional spine of the long corridor stretches east and is one storey. The plan parti is obvious: principal spaces of dining, living, bedroom and garden are on the south edge of the corridor spine; ancillary spaces north and west of it. At its eastern end is another bedroom. The three dimensional massing of the house follows almost exactly the plan parti and room size and volume. Wright referred to this functional expression as "articulation". The manner in which the patterned textile tiles ornament the building's exterior is similar to the proposed playhouse. What at first might appear the odd one out, the playhouse is in common with its contemporaries, the roof aside.

On a topographical survey map of site contour lines dated February 1924 Wright superimposed a sketch plan, nearly a parti, of the proposed Ennis house.[28] This evidence alone suggests that design began that February rather than in 1923. The circulation spine is on grade while the two storeys about the entry and garage, and therefore the motor court, are set in a very slight depression of land. In reality, except for the spine and spaces and rooms off it, the building was otherwise forced onto a long narrow site rather like a peninsula.[29] When as a result of rain storms, the dramatic tall wall that extends on the south side below motor court floor level is exposed, an anomaly is also exposed. The wall looks like it must follow the ground slope but in fact it is yards away from virgin soil and therefore retains tons of earth as backfill. This was also true on the north or street side. Nonetheless it makes a wonderful elevational composition both on paper and experientially. More than the other houses, tiles (not blocks) as used on the Ennis house appear to be dressed stone masonry.

FURTHER QUESTIONS

Questions arise when one studies the concrete blocks and tiles Wright designed for the walls of those five buildings, not their architectonics or aesthetics as applied, but things ordinary yet at the heart of notions about originality and about architectonic and structural integrity. Did Wright invent the mold system? No. As we have carefully outlined, like terra cotta, concrete cast in molds was common since Roman times, and used during Wright's life to make almost anything: fence posts, furniture, street lighting standards and gnarled park benches, and block masonry. Between 1904 and 1910 in the United States alone there was an annual average of thirty-four patents pending for concrete blocks, their chemical composition, or mold apparatuses for producing them.[30] In 1912 you could purchase clinkers of incinerated garbage to use as an aggregate to mix with Portland cement.[31] We have also established that reliable information was readily available from a variety of published sources. As well, since compressing a slurry mixture in a mold was critical, there were compaction machines that Wright or son Lloyd could have used or modified. Instead they relied on hand tamping a rather dry slurry with a block of wood and a mallet.[32]

Was the textile block and tile structural system Wright's invention? No. Blocks of terra cotta or brick or concrete, i.e. masonry, tied by iron or steel or held in position with steel rods or a expanded steel mesh across a wall's cavity (even the smallest) was a well-known and common process. In 1909 the Atlas Portland Cement Company reported a consensus important to this essay: "hollow wall [masonry] construction, which consists of tying together two comparatively thin walls with . . . steel bars at regular intervals, has been the occasion of lively interest to those studying modern building methods".[33] This supports Wright's defensive assertion, when challenged by the possibility of litigation over plagiarism, that his system was not unique. Yet he unashamedly promoted its originality, cost benefits and peerless attributes.[34] So too have others. Yet the ongoing accounts, with their lack of historical foundation or realistic understanding of the technology, have been sustained in part by an acceptance of the fiction of Original Action as presented in Wright's disembling promotions. A seldom remembered example: In April 1957, age 90, Wright was the Bernard Maybeck Lecturer in the architecture school of the University of California in Berkeley. At question time after the lecture one student asked, ". . . aren't most of your buildings relatively expensive as far as the common man is concerned?'" Wright was reluctant to answer. But the student pressed on: "I've practically never seen a low-cost thing that you or any of your students have done". Then the aged architect replied: "You haven't? Then you haven't seen the Millard House. You haven't seen the building there in Phoenix—the hotel [Arizona Biltmore]. You haven't seen the concrete block system I evolved. You haven't seen nothing!"[35]

Wright's ongoing promotions were conflated in biased and labored interpretations by assenters whose explanations, we shall soon learn, were the product of myopia and imprecision shielded by a cloying prejudice. His designs for the concrete block buildings in the 1920s, built or not but especially the four houses under discussion, have been often presented in publication, rather ineffectively in most instances but beautifully in photographs. The exception is Robert Sweeney's valuable book entitled, strangely, *Wright in Hollywood*.[36] It should be read parallel with Siry's book *Unity Temple*, those parts about concrete construction; with Smith's study of the Barnsdall commissions; with Slaton's presentation in 2001 about reinforced concrete; with the essays herein and, perhaps, with De Long's art historical notes about some of Wright's buildings of 1922 to 1932.[37]

It is clear that Wright knew best practice for concrete *in situ* and for precasting. So why did he employ mean techniques and contaminated sand, soil and aggregates? Only he could answered.

However and in spite of all words written, other serious questions remain unanswered. Biographically, practically, and perhaps philosophically, what prompted Wright to investigate concrete blocks at that moment in early 1923 and, as we have outlined, two decades after the technical developments of concrete (monolithic, reinforced, precast) had been firmly established and proven technically and economically? Practically, were Wright's block designs the unique contrivance he advanced and others lauded?, not their unique exterior sculptural surface but their interfacement, structural integrity, and physical form? More substantively and beyond plan parti, what architectural ideas infused the three dimensional appearance of the houses? Can the aesthetical characteristics of the four block houses be satisfactorily rationalized? Setting aside the aberrant unsatisfied playhouse, can interpretations heretofore be confirmed or rejected? Some people, implying nothing new since Vitruvius's elliptical thoughts or Ruskin's sentiments, will ask what were their aesthetic sources?, i.e. their precedents? Of course there are always precedents. However: "Immature poets imitate; mature poets steal". Or so said T.S. Eliot in 1920. And further, did Wright's words manage historians? The above are perhaps obvious questions. Yet they have perplexed unprejudiced art and architectural historians and knowledgeable observers, until now. Accurate answers are rather straightforward when confronted with irresistible and cooperative evidence. The search for that began in Australia.

NOTES

1   Epigraph; letter, Greenough to Emerson, 28 December 1851, his emphasis, reply 7 January 1852, in Don Gifford, editor, *The Literature of Architecture* (New York: Dutton, 1966), pp. 123-124, reply p. 125.

    The only surviving document of Wright's H.E. Brown house project, scheme 2 of 1906, is a nice perspective drawing. The exterior walls appear to be longish rectilinear blocks that could or could not have been clay or concrete masonry in compression. Years after the fact Wright recollected it was some kind of concrete block. In any event, in the absence of further information the drawing has no useful relevance to the present discussion.

2   Starr (1990), pp. 339-340.

3   Wright (1929), p. 40; Sweeney (1994), pp. 27-28.

4   Wright (1929), pp. 40-41.

5   Sweeney (1994), pp. 175-176. Millard and Wright got together again late in the 1920s but on her terms and without result. The best published illustrations of the architectural plans and textile blocks as detailed by Wright and of their installation during construction is Sweeney (1994), pp. 1-97.

6   Compare with Pfeiffer (14-24), pp. 107-110.

7   Dan la Botz, *Edward L. Doheny* (Westport: Greenwood, 1991), p. 175, among many studies of the man and the Tea Pot Dome scandal.

8   Miller quoted in Michele V. Cloonan, "Alice Millard and the Gospel of Beauty and Taste", in James P. Danky and Wayne A. Wiegand, editors, *American Women from the Nineteenth and Twentieth Centuries* (Madison: University of Wisconsin Press, 2006), p. 172, Miller's emphasis.

9   Wright (1932), p. 250, altered Wright (1943), p. 251. "Block-shell" in Wright (1943) p. 252, 314 twice, once "block shell" and "block-system". On Millard house, Wright (1932), pp. 239-250.

10 Sweeney (1994), pp. 55-56.

11 Sweeney (1994), p. 63, 246,n20; Byron Vandergrift described Storer as "older, short and heavy set", p. 246,n30. Storer's building company as a corporation was forced to suspend all business in 1931.

12 Suzanne Stephens, "Interview with Joel Silver," *The AHF Review* (Boston), 2(Spring 1993), pp. 3-4. Silver kindly allowed Robert Sweeney, Jock de Swart and me to visit the Storer house in 1985. Silver sold the house in 2002.

13 Viladas (1996), pp. 112-117.

14 Various documents on purchase and mortgage's (dated 10 March and 8 April 1924) and on marriage and teacher training, Harriet Press Freeman Archives, University of Southern California Libraries, accessed April 2012; Sweeney (1994), p. 68. Samuel Freeman died in 1980, Harriet in 1986. In 1985 Jeffrey Chusid, who as caretaker for the University of Southern California was then living in a flat under the garage, kindly escorted me around the house.

15 Interview with Byron Vandergrift in 1983 conducted by Sweeney and Charles Calvo, Sweeney (1994), p. 73.

16 <quakewrap.com/frp%20papers/seismic> accessed 5 November 2008, follow "Example case study".

17 Toker (2003), p. 114.

18 Biographical information gleaned from Sweeney (1994); Starr (1990), pp. 339-346; Smith (1992).

19 Steve Oney, "House on Haunted Hill", *Los Angeles Magazine*, 11 January 2006, accessed January 2012.

20 Smith (2005), p. 11.

21 Storrer (1993), p. 222.

22 See <www.la.curbed.com/archives/2007/08/ennis> accessed 5 October 2008. The Millard house sold for $1.3MUSD in 2000, in 2008 the asking price was $7.5M, then in 2010 it was $4.5M. In 2009, the Ennis house was for sale at $15M, in 2010 for $7.5M plus a reserve of $6M for restoration. It was purchased for $4.5M by supermarket magnate Ron Burkle who contractly had to commit to preserving the house.

23 Johnson (2005), Chapters 3 and 16.

24 Smith (1992), p. 182.

25 The nature of Wright's involvement before 1917 with promoters of progressive education and progressivism more generally is discussed in Johnson, forthcoming.

26 Drawing of the water feature in Gebhard (1993), vol. 1, plates 147, 160. Neutra shared the design experience (Smith, 1992, pp. 194-196) and his bold perspective rendering in Hines (1982), p. 80.

27 Wright's application of the square is studied in detail in Johnson (2004) and Johnson, forthcoming.

28 "Topographical Map", Beverly Hills Nurseries, "Feb. 1924"; a blueprint in color is in Pfeiffer (2002), p. 102.

29 Also see analyses by Sweeney (1994), chapter 3; Smith (2005), Storrer (1993) and, Levine (1996), passim; and so on. Wright incorrectly claimed that building permits were never issued for "the original textile block houses in California" or for many of his other buildings pre-1930, Wright (1943), p. 472.

30 Sweeney (1994), p. 216. See also articles and advertisements in concrete magazines of the period, in the magazine *American City*, 1913-1919, and in *ARecord* ca.1905-1920; the popular but rather technical Sloan (1912); and the naive proposal for concrete stud wall construction by architect E.G. Perrot in William E. Groben, "A New Type of Construction for Low-cost Housing," *American Architect*, 113(April 1918), pp. 419-424; and so forth.

31 *American City*, 9(July 1912), "The Destructor Company", p. 87.

32 Smith (2005), p. 7.

33 Atlas (1909), vol. 2, p. 6.

34 FLW to Lloyd Wright, n.d. but probably 1930, p. 1, LWPapers.

35 Meehan (1984), p. 220.

36 Sweeney (1994) and compare with Simpson (1999), Chapter 1. Other useful publications (and by no means all) during the decade of concerted research about the four houses include J. Jeff Guh, "Structural Stabilization of the Samuel Freeman House . . .", *The Quarterly Newsletter of the FLW Building Conservancy* (Scottsdale), 8(Summer 1999), pp. 1-2; Robert McCarter, "Barnsdall (Hollyhock) House", in Beth Dunlop, ed., *FLW* (London: Phaidon, 1999); Dennis J. Casey, *Art Glass Details: FLW's Hollyhock House* (Brisbane: Prairie Designs of California, 2000); Victoria Newhouse & Anthony Peres, "Joel Silver: The producer's FLW house in Los Angeles" (Storer), *Architectural Digest*, 55(April 1998), pp. 278-287ff; Suzanne Stephens & Tim Street-Porter, "FLW's La Miniatura", *Architectural Digest*, 51(December 1944), p. 102ff; Jeffrey Mark Chusid, "Concrete and light . . . Freeman house of FLW", *Antiques and Fine Art*, 7(January 1990), pp. 76-83; Pilar Viladas, "Wright in Hollywood," *House & Garden*, (February 1990), p. 78f; idem, "Invisible Reweaving," *PArch*, (November 1985), pp. 112-117; Martin Filler, "The Wright Way: FLW's La Miniatura", *House & Garden*, 160(October 1988), pp. 150-155, 231; Carol A. Crotta, "Living Wright," *California*, 13(November 1988), pp. 108-124; Renata De Fusco, "FLW Storer Residence, Hollywood," *domus* (Milan), no. 675 (Settembre 1985), pp. 92-97; Jim Tice, "LA block houses, 1921-1924," *Architectural Design* (London), 51(8/9, 1981), pp. 62-65; and so on.

Ennis House Foundation is located at <www.ennishouse.org> and includes photographs of the extensive damage caused by the 1994 Northridge earthquake and later rainstorms. The $6M restoration was complete August 2007. See also <wwww.laconservancy.org/issues/ennis>.

Pfeiffer (pp. 14-23), Pfeiffer (pp. 24-36), and Sweeney (1994) contain illustrations of—and precast blocks for—the Imperial Hotel, San Marcos-in-the-Desert, Arizona Biltmore, San Marcos Water Garden, Millard house, German warehouse, Barnsdall's Hollyhock, Phi Gamma Delta fraternity house, and a 1924 project for the University of Wisconsin.

There are hundreds of books illustrating the buildings discussed herein, but for houses by Greene & Greene, Gill, Lloyd Wright, Schindler and Wright, enjoy Tim Street-Porter, *The Los Angeles House: Decoration and Design in America's City of Style* (New York: Thames & Hudson, 1955).

37 See also Mary Jane Hamilton, "The Phi Gamma Delta fraternity house", and Paul E. Sprague, "The Marshall Erdman prefabricated buildings", those in concrete, Sprague (1990), pp. 67-76 and 151-161.

3

The Taylors and the Griffins

To "invent" is usually defined something like this: to originate as a product of one's own contrivance, or to produce or create with the imagination, or to make up or fabricate as something merely fictitious or false. So two synonyms might be "to create" and "to fantasize". A meaning more relevant to this essay is: to devise something new, as by ingenuity. Devise?: to order or arrange the plan of something; to think out; to plan; even to contrive. An invention in patent law is the conception of an idea and the means or apparatus by which the result is obtained. For our purposes, therefore, we are concerned with originality, the beginning of something, origins, roots, and the means to realization.

THE TAYLORS

Australians George Augustine Taylor and wife Florence Mary Taylor were founders, owners and editors of *Building*, a building construction magazine published in Sydney, New South Wales, under various subtitles from 1907 to 1972. Their crazy relationship with the American architecture and city planning team of Walter Burley Griffin and Marion Mahony Griffin is one key to understanding a source of Wright's initial block design for Millard. Recognition of that key was mooted in 1970 and updated 1973.[1] Now, with new evidence and fresh analysis derived from diverse sources, the picture is much clearer, confirmation unambiguous.

The Taylors were hyperactive in work, recreation and community service. As a young man George learned about building trades at Sydney Technical College and in the 1890s turned to cartooning for the Sydney *Bulletin* and London's *Punch*. He mingled within avant garde artistic and literary circles but around 1900 shifted interests and became fascinated by emerging technologies and spiritualism. The first Australian to fly an unpowered biplane of his own construction (on 5 December 1909), an inventor, flautist and singer he was also a zealous nationalist.

Florence's parents the Parsons emigrated from Bristol, England, when she was age three. In late teens she obtained clerical work in the office of Sydney architect Francis Ernest Stowe before beginning articles with architect Edward Skelton Garton around 1901. She studied evenings at Sydney Technical College's 1900-1904 while continuing with Garton until 1906. That year she began employment with J. Burcham Clamp, an architect with a casual interest in town planning and in British Arts and Crafts. (Perhaps with Florence's encouragement the Griffins and Clamp were partners on a few Sydney architectural projects during 1915.) In 1907 Florence became the first Australian woman to qualify as an architect, a fact the Royal Australian Institute of Architects reluctantly accepted in 1923 after sixteen years of chauvinism.

After marriage in 1907 the Taylors soon became involved with dozens of professional, art, theater and civic organizations. They also set about publishing magazines and books devoted to favorite subjects: building construction, aviation, soldiery, wireless (radio), mechanical engineering, government, architecture, and town planning: manly subjects. While in print the Taylors might praise, they were more inclined to probe, criticize, needle and abuse governments, professions, organizations and individuals. Their first and flagship serial publication was *Building* magazine, begun in 1907 with a loan from the Stowe.[2] Through *Building* they enthusiastically encouraged the government sponsored international competition of 1911-1912 to design an Australian capital city. As editors they passionately supported he Griffins and their winning design for the layout of what became Canberra. But the Taylor's treatment of Walter and Marion would unexpectedly and bitterly change in 1915.

Walter was born 1876 in Maywood, Illinois and obtained an architecture degree from the University of Illinois in 1899. He worked for Wright from sometime in 1901 to January 1906 when he began to independently practice landscape and building design. He was joined by wife Marion in a practice that lasted from 1911 into 1937. Marion was born in 1871 and obtained an architecture degree from Massachusetts Institute of Technology (MIT) in 1894, the second American woman to do so and she received an Illinois license to practice architecture in 1898. From 1895 to 1911 she worked in a few Chicago offices but mainly for Wright. Marion and Walter married in 1911 and began working on a design for Canberra. As a result of winning the competition Walter was invited to Australia in August 1913. After a few weeks he accepted a contract with the federal government to oversee the design, landscape, and construction of Canberra. In *Building* Taylors publicly and joyously received Walter in 1913 saying,

> [Burley Griffin] speaks as if inspired by a love, and not the least moved by a regret, for a young nation of glorious possibilities upon which the mildew of medievalism has already settled....
>
> We welcome Walter Burl[e]y Griffin as an ally in our fight for a progressive spirit in Australian architecture.[3]

In April 1914 the Taylors' offered another welcome: "We hail Frank Lloyd Wright and his disciples [Walter and Marion Griffin] as great missionaries in the cause of Architecture". Below this accolade the Taylors reprinted the text of Wright's important theoretical treatise that had appeared in 1908 in the New York *Architectural Record* entitled "In the cause of architecture." The Taylors' reprint was illustrated with only a couple of Wright's buildings and, unlike the original, Walter's.[4] The Griffin entourage then arrived in Melbourne on 14 May 1914.

That generous reception and eulogistic tone continued in George's optimistic book entitled *Town Planning for Australia*, self-published in late 1914. Drawing in part on information the Taylors gleaned on a North American tour, in outline it informed Australians of the progress of city planning in England, Europe, Canada, the United States, and familiarized readers with a number of Griffin's housing and other architectural works, all amply illustrated.[5] A better informed and more useful city planning document, however, was a report by Melbourne architect J.C. Morrell who had travelled on a government sponsored fact finding tour to Britain and North America that same year: 1914.[6]

IN CHICAGO

The Taylors, Sydney builder Richard H. Pearce and engineer Ernest Stowe, who was Florence's first employer and the son of close friends of George's family, travelled together to North America during June and into August 1914.[7] Landing at Vancouver, B.C. they rolled south by train to Los Angeles. From there and through the desert Southwest their train swung north to Chicago, "the Mecca of their ten thousand miles pilgrimage." The Taylors carried with them introductions written by the Griffins to former Chicago architect Myron Hunt then in Los Angeles, and in Chicago, to the Griffins' social and professional friends. Among them were Walter and Marion's parents, architects William Gray Purcell, Barry Byrne (former Wright employee and soon to be Griffins' U.S. partner), Dwight Heald Perkins (Marion's cousin and her and Walter's former employer) and wife and author Lucy Perkins, Wright his former employer, and theorist Louis Sullivan.[8]

In tow by the Perkinses or Walter's father George, the Taylors visited a Chicago Architectural Club exhibit mounted at the Art Institute, Jane Addams and Hull-House settlement, the Women's Club and Women's City Club, and the all-male City Club where Perkins, Walter and George were active members. Walter was then chairman of the club's City Planning Committee. Each club organization was actively promoting better housing, city play grounds, municipal parks, forest preservation, Negro rights, urban cleanliness, woman's rights, and other liberal issues. The Taylors professed an interest in—if not possessing a knowledge of— some of those issues.

George Taylor gave a long talk on Australian architecture to the City Club where he referred to Walter as "the greatest town planner." However, the impressionable Taylor was critical of Chicago's housing, its slums (those "mean streets"), and workers' conditions.[9] Emboldened by and taking a lead from their progressive-minded hosts, George reckoned that Daniel Burnham's 1909 Chicago city plan was too "aristocratic to touch the poor people in the slums!."[10] He praised those local activists who opposed it not only in economical terms but on moral and ethical grounds.

During an interview, Sullivan gave the Taylors a lithographic reproduction of a perspective drawing of the main portal of his and Dankmar Adler's Transportation Building, designed and built for the Chicago World's Colombian Exposition of 1893. The print was autographed by Sullivan:

> *To Capt Geo. A. Taylor*
> *with Kindest wishes,*
> *Louis H. Sullivan,*
> *Chicago July 29, 1914*[11]

(George was commissioned in the Australian Intelligence Corps in 1912.[12]) During 1914 the Taylors serialized in Building an essay by Sullivan entitled "What is Architecture?"[13] Before departing Sydney the Taylors learned that it was Sullivan's theoretical utterances that intellectually sustained the young, bright and eager Griffins.

American architects gave the Taylors and Stowe many photographs of buildings and streetscapes. Wright also gave them a copy of the then expensive now valuable two volume folio publication of 1910, *Ausgefürte Bauten und Entwürfe von Frank Lloyd Wright*, published by Wasmuth in Berlin. It was inscribed,

> *To Captain Taylor*
> *from his friend and admirer*
> *August 6, 1914*
> *Frank Lloyd Wright*[14]

At the recently opened Midway Gardens, a beer-garden night-club or sorts designed by Wright during 1913, the Taylors and Stowe enjoyed "music, architecture and gastronomy." It was a "fantastic" building, the Taylors claimed, the "impulse of its clever architect." Yet, while in high praise of Sullivan, Perkins and a few others, they now believed Wright was too preoccupied with the unconventional. He was, they observed with odd logic,

> *a clever Chicagoan [who] embraced the Sullivan idea, but he lacked the strength of personal character needed to give stability to his achievement.... To achieve notoriety he sometimes aims at [architectural] eccentricity—even at the expense of sincerity of construction.*[15]

They also repeated Wright's gripes and recriminations, such as; "Some of his [Wright's] pupils [employees] who are now scattered, are reproducing these [architectural] freaks and prattling the Sullivan phrase: "Form follows function," without apparently knowing what it means."[16] Yet at no time then or after did the Taylors reveal they understood its meaning. Indeed, what they did say about architecture betrayed some ignorance if not intellectual innocence. On the other hand they believed that the likable Perkins knew its meaning: "He [Perkins] is a man with the Sullivan ideal, that in architecture 'form should follow function', but unlike the Wright school of constructional eccentrics he does not prattle the phrase, but practices what it preaches."[17] The article illustrated a house by Perkins executed in an idiom only reminiscent of the then popular English Arts and Crafts as made popular by the architect C.F.A. Voysey.

The Taylors' remarks about Wright and back handidly the Griffins continued unabated:

> He [Wright] notes the passing away of the freak style of architecture that he endeavored to foster, and for that he blames his pupils [sic], a band of young men and women he employed in his office. Some, noting that the new idea of architectural design seemed rather easy and did not require any of that knowledge of the principles of construction calling for many years of close study to master, cut out from Wright's office. And some of them took Wright's clients! They are doing works that Wright pathetically describes as "half-baked imitative designs, fictitious semblances pretentiously put forward in the name of a cause."[18]

"Freak" was one of the Taylors' favorite architectural terms. Those comments were paraphrases gathered not only from personal discussions with other Chicago identities but taken from a copy of Wright's just published, at times bitter essay that had appeared May 1914 in the *Architectural Record* entitled "In the cause of architecture, Second paper."[19] Still with the Griffins in mind the Taylors selected a few lines from that article.

> My disciples or pupils, [Wright said] be they artists, neophytes or brokers! I dread to see the types I have worked with.... cheapened or be fooled by senseless changes, robbed of quality and distinction, dead forms or grinning originalities for the sake of originality, an endless string of hacked carcasses. ...[20]

About a year later and after certain events we will soon outline, the Taylors now believed that "Wright and his disciples [had] . . . drifted from the common sense in construction. They had released their grip of the guiding ropes and so became lost in a labyrinth of eccentricity and fake ideals."[21] Ropes? Those comments were written while the Taylors were in the U.S. and had them relayed back to Sydney for publication in *Building*. Surprisingly, the Taylors had not as yet met Wright! When they finally did so that August in Chicago they had difficulties appraising the garrulous out spoken American: they found him to be

> an entertaining genius, and a good-natured fellow. He has the long hair, the love for "uttering platitudes in stained-glass attitudes" [?], the "morbid affectation born of love of admiration," of the old-time Oscar-Wilde school....

Moderately concerned with social issues, George, Florence, and Ernest were rigidly conservative. Within their political and social nexus "the Taylors were right-wing business-people who denounced unions and welfare".[22] As bellicose conservatives they easily became enamored with fascism as a militant share of nationalism and patriotism.

Traveling on to Buffalo, New York, the Taylors and Stowe attended Wright's "bleak-looking" and "mausoleum-like" (their description) Larkin Administration Building of 1902-1906, "designed by a 'soulless engineer,'" they moaned.[23] On returning to Chicago that August they finally met Wright. He pulled out the Larkin drawings and photographs. When queried he confessed that "yes," the building did have a . . .

> *forbidding aspect . . . [but] I wished to suggest [Wright said,] that the work of the office [i.e. Larkin's personnel] is a thing apart from the sordid world around it: that the quietness and calm necessary to every office interior should be preserved by giving the building an exterior appearance of "keep out, this is no place for the curious—move on!"*[24]

After experiencing his buildings and then listening to Wright, Florence retreated further.

> *We have laws, [she said, to check] any daring break from public opinion. Convention is often the law of the majority, and the man who would be original enough to dare such a thing as that might dare to break the conventions in other things and be dangerous to somebody.*[25]

Dangerous. It is clear the Taylors found leftist Wright to be a psychological curiosity, his architecture aberrant and, finally, untenable. These were findings they would soon attach with barbs to the Griffins. Moreover, there can be no doubt there were discussions with Wright and with all others (as we shall learn) who held opinions about the Chicago expatriates then living in Australia, just as Wright learned much about his former employees in Australia.

The "animus" between Wright and the Griffins was to prejudice the senior architect's view of the information he gleaned from the Australians. Wright even brought Frank Byrne into the fray. In an attempt to calm his former boss, Byrne told him to "Let it rest—your greatness is not served in that way."[26] Byrne had worked for Wright and was in 1914 and 1915 supervising the Griffins' commissions in America.

The threat of a European war that July thwarted the Taylors' and Stowe's plans to travel on Atlantic seas to "home," to England's United Kingdom. So in August the Taylors returned to Los Angeles via Canada, Chicago and then Minneapolis. There they enjoyed a full day with Purcell, a trusted friend of Wright's and perhaps the Griffins best friend in Chicago's architectural orbit. Purcell remembered the occasion well, finding the visitors from down under "an exceedingly low cast" lot. Further, he said:

> *One political editor, his wife and an associate came to America to rake up dirt about Griffin and thereby discredit him in order to secure control of the capital city project. The party came to see me . . . with voluble but obviously insincere expressions of friendship for Griffin. I was fortunately quick to sense the situation and was able to cancel with facts the hypothesis of villainy with which they were trying to tag Griffin . . . while they did not get one shred of concrete evidence of incompetency against Griffin in this country, they continued to send back [to*

Minneapolis] a sequence of scurrilous newspaper correspondence together with articles in their own weekly.[27]

We have mentioned that during 1914 and 1915 some of the Taylor's travel memoirs had been relayed to Sydney and serialized in *Building*. The articles were accompanied by examples from the many photographs received from American architects and city engineers. Together with additional material, if fewer pictures but retaining those negative comments about Wright and by obvious implication the Griffins, George produced a book about their tour. He entitled it *"There!" A pilgrimage of pleasure* and it appeared in 1916, again self-published. Buildings by Purcell, Perkins, Wright, Sullivan, and many other architects were illustrated, but none by the Griffins. The turn against the Americans had become perversely opposite to their reception in 1913.

HOSTILITIES

There are two reasons for the Taylors' publicly expressed dislike of the Griffins. Inspired by the positive results of the Canberra competition and of the various social settlements, womens' and civic and other improvement clubs and associations witnessed in America, they were prompted into verbal action. In October 1913, before departing for the U.S., along with English emigrant architect John Sulman, Frederick Stowe and son Ernest and others including politicians and architects, the Taylors helped found the Town Planning Association (TPA) of New South Wales.[28] During early 1914 Marion, Genevieve (Walter's sister) and husband architect Roy Lippincott (both had travelled to Australia with the Griffins), set up a Sydney office to attend to private architectural and city planing commissions forthcoming in New South Wales. Walter remained in Melbourne to handle local commissions and direct planning and construction of Canberra.[29] He was, however, co-opted to TPA's council. Marion joined the Women's Section as an elected vice-president. The section was formed in March 1915 by Florence (president) and the wives of politically and commercially influential men. Their rather ill-defined platform related only slightly to those of the varied women's and other welfare clubs Florence had witnessed in American cities.[30] The Women's Section quickly attracted one hundred members who were concerned with presenting the "human side" of planning as well as issues traditional to wifely issues.[31]

Impressed by progressive developments in ameliorating the effects of an urbanizing America, and in support of TPA, the Taylors began a new journal in 1915 entitled *Town Planning and Housing Review*. It presented thoughts and ideas about cities and outlined programs of newly formed town planning associations in each Australian state, especially advances in South Australia through the town planning efforts of the newly appointed state planner, the respected and well-traveled Charles Reade. But emphasis was on New South Wales, home territory. Sadly, without advertisers TPHR ceased publication after six months.

However much the separation of women and men irritated, Marion put all effort into strengthening the womens' section. In late 1914 the Taylors invited Marion to adjudicate a modest suburban landscape design competition sponsored by *Building* for "a garden of one acre." Her judgement was published in January 1915.[32] That politeness ended abruptly. In a February article Florence referred to "American ladies" living in Australia (obviously Marion) as "invader[s] from foreign countries."[33] It was more than a hint of things to come. During the following months or May to November 1915, the Griffin/Taylor relationship irreconcilably soured, a tussle exposed in print by the Taylors.

Most TPA men were opposed to full female participation in any form. Chairman was the arch conservative John Sulman and he quickly "pointed out" the genetic fact of male superiority. Incensed, Florence, Marion and some other women resigned their elected offices. George ordered Florence to withdraw her action. She in turn accused Marion of upsetting "everything." City planning historian Robert Freestone put it this way. Marion simply could not accept the position put by a "very suave" Sulman that women were incapable of organizing a special section. Because of Florence's capitulation, as it were, Marion reactively described her "a pathetic figure." The women's section "soon fell apart," Freestone has said, "divided initially by personality conflicts, conflicting ideological agendas and ultimately quashed by the power of the all-male executive."[34] Ernest Stowe, TPA treasurer, "whose patronizing comments about the scope of women's [daily] work had drawn snickers from several ladies present," decided unilaterally to cancel Marion's membership.[35] Marion's recollection was personal, incisive, negative. Under the heading "MMG, Town Planning Association 1915" she wrote in a memoir:

> *The first year in Australia was full indeed with doors opening and doors slamming in our faces. . . .*
>
> *And curiously enough Griffin's fight with the Government was reflected in my battles in private life.*
>
> *The publisher of a magazine[,] Mr and Mrs Taylor, and their pal Mr [Ernest] Stow[e] who had tied up with us from the first days, called me into their [Sydney] office and told me that from now on [Walt] Griffin was to do what they told him to do in Federal Capital matters etc. . . . A number of very fine and very capable women lined up with me in the democratizing of Sydney's Town Planning Association.*[36]

But they failed. The Taylors and their influential supporters were barely liberal in word, certainly conservative in action. In their view Marion represented a "dangerous," a progressive, foreign and unnecessary reformist element.

Mid-1915 the Taylors called for Griffin's removal as Director of Design and Construction for Canberra. They alleged he had proven to be incapable "of bringing the development [of Canberra] to a successful issue:"[37] a charge wholly incorrect as events proved. George Taylor and Stowe suggested the establishment of an apolitical commission to superintend Canberra's construction. Beginning December 1916 scurrilous and derisive words appeared on *Building's* pages, all apparently designed to discredit Walter's efforts for Canberra and discourage potential independently acquired architectural and town planning commissions. Historians Freestone and Hanna's evaluation is pertinent:

> While united in their interests in the built environment, sharing a similar disdain for big government, and equally [?] energetic and passionate about social and political causes, the Taylors and the Griffins were very different people. The Taylors were both of working class stock, poorly educated, materialistic, right wing and interested in social prominence and business success. The Griffins were middle class, well educated, spiritually included, left leaning, and interested in art and ideas.[38]

Armed with information gained in talks with a touchy and disagreeable Wright in 1914, the Taylors went so far as to question Walter's credentials to a Federal Parliamentary Commission investigating the administration of his efforts for Canberra. In progress 1916 into 1917 the commission was formed as a result of some disgruntled bureaucrats architects and engineers haranguing federal politicians,. The Taylors claimed unprofessional acts and curious decisions relevant to the progress of work. The commissioner easily found in Walter's favor.[39] But the knife of discredit had cut deeply.

A few years later the Taylors renewed their attack on Wright. One highlight was a reprint in *Building* of an essay by the irascible California architect Louis Mullgardt whom the Taylors may have met in Sydney when on his world tour 1922-1923. After visiting Wright's Imperial Hotel in Tokyo in 1922, Mullgardt sent a manuscript post haste (he said) from Beijing, China, to the San Francisco professional magazine *Architect & Engineer*. Published in November it was a scathing, at times irrational attack on Wright personally and a condemnation of the hotel's exterior and interior appearance. In one instance Mullgardt railed:

> it is a monstrous thing prehistoric in plan, design structure, decoration and the state of decay....
>
> [O]ne sees a fortress of buff brick and terra cotta; every facade has been laminated and lambasted with a stone of exceeding rottenness, which has been much carved with patterns of Yucatanese, Aztec and Navajo piffle.[40]

Wright's response, entitled "He who gets slapped," was published only in part and only in the popular English language newspaper, the Japan *Advertizer* of 7 February 1923. It seems that Mullgardt sent Taylor a copy of his manuscript for it was printed in *Building* also that November. Editorially George likened the Tokyo hotel to a "Griffinesque effusion" designed for an office building facade built in central Melbourne.[41]

The San Francisco publication also printed a letter by the Canadian expatriate American architect F.W. Fitzpatrick, friendly host in Washington, D.C. to the Taylors during their 1914 visit. He thereafter was a sometime correspondent to *Building* and on one occasion wrote in support of Mullgardt's testy article. Mullgardt and Fitzpatrick gave slight credit to Sullivan for "doing something original," as Fitzpatrick put it, but his "disciples Wright has sinned the most and the worst. And this last sin seems the most sinful of all past sins." Major portions of Fitzpatrick's letter were republished in *Building*.[42]

We should note that it was probably in the early 1920s that the Taylors became good friends with and generally supported a right-wing, protofascist Sydneysiders of some notoriety and others of similar mind including a fascistic group called the New Guard.[43] The Taylors' "progressive spirit" was short-lived and succumbed inherently to the less favorable aspect of conservative ideology.

Another reason for the Taylors' turn against the Griffins and perhaps Wright was the neutral position of the United States during the war years of 1914-1917. Walter and Marion had apparently expressed opposition to American participation in Europe's family war, a war to which the English colony of Australia was unreservedly committed. As aggrieved Anglophiles George and Florence became vocally anti-American. Perhaps their first public outcry against the Griffin entourage was in April 1915 when George wrote a response to a published letter to the editor from Walter's brother-in-law, Roy Lippincott ("an American" George warned). Like the Griffins and Wright, Lippincott was a pacifist and protested the publication in 1915 of what was to him an anti-Christian, pro-war cartoon drawn by George. Lippincott's letter enraged the cartoonist. In a *Building* editorial Taylor zealously said much against Lippincott's views and assailed "the so-called U.S.A. peace policy," saying it was "based on greed and fear:" profiteering and fear, perhaps of the government to recruit an army. Although the war had not begun, he said that while in the U.S. during 1914 he witnessed Negroes being ostracized in peace but when faced with raising an army the government advertised "Colored men wanted," that, as Taylor put it, "Sambo is a brudder, sure."[44]

In truth the proportion of Negroes in the American armed forces 1917-1918 was about equivalent to the national population, a segregated one in ten. And we know something of George's—and most of Australia's—prejudice. On one occasion he said, "The white race being superior . . . [will] guide the future of humanity."[45] It seems that contrary to British sensibilities and practice at home, George and Florence agreed with the Australian federal government's official "white Australia " policy, one openly expressed and to continue into the 1960s.

Taylor's criticism of Wright and the Griffin camp insinuated them to be deviants, not conventional, perhaps heretical, that in difficult times for the mother country this was unacceptable, disloyal. In the Griffins' case, the example was a series of comments by George in 1915 about some buildings for which Walter was responsible at the newly established Military College at Duntroon in Canberra. One article was headlined:

Canberra's Comedy
Queer Buildings Proposed
An Insult to Australian Architecture

"Abortions" and "atrocities" were to be "the first buildings of our new Capital!," Taylor exclaimed. They "were designed by W.B. Griffin, who Australia does not claim as an Australian and who certainly does not repudiate his American origin."[46] As designed, the proposed storage buildings were small, simple brick boxes nicely detailed.

Further, the Taylors rather mischievously suggested German tendencies in the Griffins' architecture. Historian James Weirick discovered that that "theme"...

> was taken up [by] the [nation's] Governor-General, Sir Ronald Ferguson, who... in 1918 observed that "a banal, clumsy German architecture, typical of German 'Kultur' had recently been disseminated over a suffering earth, and was reaching Australia via America."[47]

Ferguson was English, appointed by the King, not an Australian. Moreover, the Australian government's Minister of Home Affairs, Mr. Archibald, referred to Walter as being "a Yankee bounder:" a noisily ill-bred American.[48] Marion remembered those desperately uncomfortable years. In her experience "a foreigner" in Australia was, she said, "to be feared, to be hated, to be despised ... [one] whose honesty, intelligence, industry are things to be deadened.... The whole community unites to hound, to cheat, to defame the foreigner...."[49] In a 1915 letter Marion continued expressing discomfort:

> There is a very strong anti-American feeling here .... There is no chance here for any foreigner (and any new-comer is considered to be a foreigner) to prosper in any field, trade or profession ... the freezing out of strangers .... It is a nation of pessimists full of fears.[50]

Composed of colonies set up as thoroughly British enclaves—nearly counties—in the antipodes, it was difficult times for the fledgling, underpopulated white Australians. Yet, without hesitation they went to the defense of King and "home" counties when that war between royal families began August 1914.

During March and April of 1914 Walter and Marion travelled to Europe to exhibit their architectural designs in Paris composed mainly of Marion's wonderful ink and guache drawings. Formally they were officially there to select a jury of architects to adjudicate an international design competition the Griffins quite properly were organizing for a new federal parliament house. In England they secured John J. Burnet, president of the Royal Institute of British Architects; in France the beaux-arts architect, teacher and town planner Victor Lalous; in Austria the venerable architect and teacher Otto Wagner; in the U.S. an ailing Louis Sullivan (much favored over Wright); and the talented Australian architect George Temple-Poole in Perth.[51] Only months after the outbreak of Europe's war the competition was suspended by the government. It was reopened in August 1916 to only British, Australian, New Zealander, South African and other competitors in colonial outposts. Then it was officially postponed that November. Four years later, December 1920, the federal government abolished Griffin's position of Director of Design and Construction of Canberra and immediately cancelled the competition.

The Taylor's intemperate hostility continued through 1927, the year a new Parliament House (designed locally) was opened in Canberra and well after the federal government had succumbed to a multitude of direct and indirect negative pressures.

With George's untimely death in 1928, victimization ceased. George, Florence, Ernest and their accomplices had lost 1915-1919 skirmishes but won the war.

CONSPIRACY

In late 1919 two of Taylor's friends, Stowe and an unnamed friend, travelled to America and included stops in Minneapolis, Chicago, and Los Angeles. When informed of their impending visit to Minneapolis that July, Walter wrote a worried note to "Dear Will Gray" Purcell, then practicing in the Minnesota city. Purcell, then in Chicago, was asked to take "charge" of Wright's architectural practice but he wisely declined. Wright did "appreciate" Purcell's' "reasons to not undertaking to carryon the work here . . . ;" here being Wright' office.[52] In 1909 Wright was organizing an extended trip to Europe with Mrs Mamah Cheney. Anyway, in part Griffin said (in 1919) to his Minneapolis colleague and dear friend:

> *I hope you have been very careful in statements you may have made to George Taylor's partner[,] whoever he may be this time. As a result of their former visit [in 1914] to America, Taylor and his partners [presumably Pearce and Stowe] to whom I introduced my best friends at home [America], warned Prime Minister Hughes [in 1916] of my disreputable character in America in order that his government should not make the mistake of reengaging me, which they did however [in 1917].*
>
> *Taylor's opinion of you, I take it, must be included in Stowe's written statement that I belonged in America to a "long-hair, bow-tie, cigarette-smoking[,] absinthe-drinking cult"[.] . . .*
>
> *However, Taylor is not an isolated example[,] I am sorry to say, for it is surprising how much time men in a small community like this [Australia] can find for solely destructive enterprise. . . .*[53]

Stowe and the Taylors' had visited Purcell in 1914 and remembered the Australians' very well, finding the Australians "an exceedingly low cast lot". He explained:

> *One political editor [George], his wife and an associate [Stowe] came to America to rake up dirt about Griffin and thereby discredit him in order to secure control of the capital city project. This party came to see me in Minneapolis with voluble but obviously insincere expressions of friendship for Griffin. I was fortunately quick to sense the situation and was able to cancel with facts the hypothesis of villainy with which they were trying to tag Griffin . . . while they did not get one shred of concrete evidence of incompetency against Griffin in this country, they continued to send back [to Minneapolis] a sequence of scurrilous newspaper correspondence together with articles in their own weekly.*[54]

A mechanical engineer who also acted as an architect, Stowe lectured in surveying at Sydney Technical College. He wrote a series of short rather peculiar articles for *Building* entitled "City Design and Town Planning" that ran intermittently 1917 to 1920. He opposed the Garden City concept of town planning and the garden suburb then in vogue and other radical or "socialist" endeavors, and the Griffins. In June 1915 Walter gave a talk to the Architectural Students' Society of the Royal Victorian Institute of Architects in Melbourne. Within the presentation was a theoretical proposal for a modest "One-Roomed House." A condensed and edited version of the talk that related to the one-room was published in the Melbourne *Herald* newspaper. As the Taylor's respondent for *Building* "Major" Stowe put down that version in rather mean and cynical terms, finding the idea "utterly abhorrent to our people" and a "pernicious American method of housing poorer classes." Oddly, he likened the potential lack of privacy to the Japanese "practice of the sexes bathing together in quite a nude state."[55] The one-room design was realized in three tidy cottages constructed of knitlock during 1920 to 1922.

In any event there is a nice causal fit between George Taylor's diatribes, his and Stowe's obsessionally expressed biases the observations of Walter, Marion and Purcell. Of course Walter and George had friends willing to pass on spicy, spiteful information about the other. The evidence affirms the view that again George and Florence wanted to embarrass or discredit the Griffins professionally. The Griffins recoiled but uttered no defence in response, at least not in print.

It was Stowe who, in 1919 and on yet another of his frequent trips to America,[56] passed on to the Wrights and employee Rudolph Schindler, information about the Griffins' new "knitlock" concrete block structural system. Of this Marion had no doubt when in 1947, then resident in Chicago, she wrote to Purcell, then living in Pasadena, California about the collusive incident, saying that,

> A couple of Walt's enemies in Australia came over here [America] on a jaunt and made contact with Wright and told him about Walt's Knitlock concrete buildings. Immediately to establish his claim[,] Wright built a tile house [for Millard] of square concrete blocks as thick as the usual masonry walls and of course with no interlock.— ...
>
> His [Walter's] invention of the knitlock concrete interlocking block construction was perhaps the last work [word?] in space economy construction. It could be used with plastics as well as with concrete.[57]

The possibility to cast a plastic material is interesting, bakelite then in use. Marion repeated the same information in similar language in another letter to Purcell of 1949.[58] In memoirs written intermittently during 1939 to 1949 only her phrasing was different: "[Wright] listened to the knitlock story and shortly after built this structure [the Millard house] in California. He was always quick on the uptake".[59] She then said that Wright, a man with a "malicious vanity,"[60] used "an Australian who contacted him on a trip through America and thence forth forwarded to him [Wright] what he could get hold of of Griffin's work as for instance an imitation knitlock house in California."[61] In other words after returning to Australia the Taylors and Stowe mailed the information sometime at the end of 1919 to Wright's office. It should be noted that knitlock came to public attention in Australia only after Stowe's visit to America. It was first publicly demonstrated in September 1919 at the Royal Melbourne Show, a state agricultural fair. Perhaps knitlock literature prepared for that show was also sent to Schindler and Wright Junior.

If we accept the words of Wright's other professional associates like Barry Byrne (even if thought rather untrustworthy by the Griffins[62]), or Purcell or son Lloyd or architects George Elmslie or Robert Spencer Jr. or Schindler, in matters related to Wright's biography and architecture, then Marion's must be considered to possess equal correctness and authority, perhaps greater. After all, she was employed by Wright off and on (mostly) from 1895 to 1911, he enjoyed her trust and repartee, and she often attended to his children at their mother's request. In 1908, in the *only* public praise of an employee before 1935, Wright described Marion as "a capable assistant for eleven years."[63]

No doubt Marion obtained most of the information and gossip about the exploits of Taylor and Stowe from friends in America and Australia. Moreover, she and Walter had visited Chicago and Los Angeles in 1924-1925. There they met again those who had worked for Wright, including son Lloyd and Schindler (Wright was at home in Spring Green, Wisconsin), clients Freeman and Millard, and maybe the Ennises. Those individuals would have described to the Griffins what transpired during the period 1919 in to 1923.

It is true that Marion had grown to dislike Wright, mainly as a witness of his conceit and the unfair treatment of wife Catherine and of Walter and other employees. Equally, Wright had come to dislike Walter, mainly because he demanded repayment of a loan and wages due, cash rather than the Japanese prints received.

One measure of Wright's antipathy is found in a comment by Schindler, an Austrian architect who in 1914 emigrated to the U.S., worked in Chicago, and later was employed by Wright in Chicago and Los Angeles from 1918 to late 1921. Schindler had not personally met Walter but received biased information about him from Wright and his architect sons John and Lloyd who on occasion had worked beside Schindler. Using that skewed intelligence, in March 1921 Schindler wrote to his friend in Vienna, the architect Richard Neutra, saying in reply to a query about Walter, that

> Burley Griffin [sic] was with Wright only a short time. He copied his [Wright's architectural] mannerisms and as soon as he was on his own in Australia he failed. From what I hear [we emphasize] he is a doubtful character who by sheer chance won the competition (the conditions were such that most of the architects refused to participate). Anyway, the buildings he is doing in Australia are just empty bluff.[64]

Schindler's reply is an example of paraphrased Wrightian language and only the part about settling in Australia is correct. Walter's employment with Wright from late 1901 to January 1906 was two years longer than Schindler's. Also, in America and by friends in Australia, Griffin was commonly addressed as Walt or Walter. "Burley Griffin" is typical Australian usage (once "Burleigh") and probably disliked by Walter and Marion. (The principal lake in Canberra is incorrectly called "Burley

Griffin": it should be Lake Griffin.) Anyway, Schindler's comments were no doubt *also* swayed by tales put in 1919 by Sydneysider Stowe, a visitor to Wright's office.

+++

In architecture Walter Griffin was—in fact proudly admits to being—persuaded by Louis Sullivan's stirring words. He was also and naturally influenced by Wright's exuberant talent. But around 1910 Walter developed an independent architectural voice and style. In city planning, however, the Griffins' designs were theoretically and practically far in advance of many practitioners in America, certainly of those in Australia and of Wright, as tantalizing as his few community planning projects may have been.[65] Walter and Marion were hardly the "half-baked imitative" designers Wright had intolerantly represented.

While personal feelings may color how information is presented, the information itself can be tested and verified, or not. In this case a variety of disparate sources assemble in comfortable agreement. The above evidence introduces a tangle of reasons for the Taylors' and Stowe's to engage in a relatively minor if rather spiteful act of industrial espionage (shall we call it?), and its acceptance by the Wrights. While written testimony sufficiently supports the claim of conspiracy leading to abuse, what of the physical and architectural evidence? Did Wright borrow the knitlock design? If so, how did he avoid a charge of plagiarism?

NOTES

1. Donald Leslie Johnson, "Notes on W.B. Griffin's 'Knitlock' and His Architectural Projects for Canberra," *JSAH*, 29(May 1970), pp. 188-193; idem, "Walter Burley Griffin: An Expatriate Planner, *Journal of the American institute of Planners,* 39(September 1973), pp. 326-336; Johnson (1977), pp. 133-115.

2. For biographies of George and Florence Taylor see Freestone/Hanna, Chapters 1 and 2; on Stowe's architecture see Berith Park house at <wahroonga.org/berith_park.htm> accessed 28.1.09.

3. Editorial introduction to Walter Burley Griffin, "Architecture and Democracy" *Building* (Sydney), 13(October 1913), p. 64a; c.f. Griffin (2008). On the Taylors see also Hanna (2002), p. 31ff; Roe (1984), Chapter 7; Freestone/Park 2009); on George see Michael Roe, *Australian Dictionary of Biography 1891-1931*, vol. 12, (Melbourne: University of Melbourne Press, 1990), and on Florence, Christa Ludlow, ibid; on George see Geoffrey Serle, *Dictionary of Australian Biography* (Sydney, 1949); see also J.M. Giles, *Some chapters in the life of George Augustine Taylor* (Sydney: Building, 1957); and on Florence, Stephen/McNamara/Goad (2006), pp. 221-223; J.M. Giles, *50 years of town planning with Florence M. Taylor* (Sydney: Building, 1957); Julie Willis, "Designing Comfortable Homes, Women in Little Known Jobs," *Fabrications*, 10(August 1999), pp. 47-48; Freestone(1995), pp. 263-267; Julie Willis and Bronwyn Hanna, *Women Architects in Australia 1900-1950* (Canberra: Royal Australian Institute of Architects, 2001), p. 6ff.

4. George Taylor, introduction to Frank Lloyd Wright, "How the Movement for Democratic Architecture is developing in America," *Building* (Sydney), 13(April 1914), p. 67, a slightly abridged reprint without the very important illustrations of original Wright (1908).

5   Taylor (1914a).

6   J.C. Morrell, *Town Planning Report...* (Melbourne: Government Printer, 1915); a slightly abridged reprint in the Taylor-owned *TP&H*, 1(29 November 1915), pp. 1-33, and, 1(31 December 1915), pp. 1-32.

7   "During the year" of 1914, Taylor and Stowe "visited America," see "Town Planning and Housing," *TP&H*, 1(April 1915), p. 5. On Stowe see Obituary, *Sydney Morning Herald*, 21 July 1936, p. 8b.

8   Taylor (1914b), p. 71.

9   Taylor (1914b), p. 74, with thanks to Christopher Vernon for bringing this information to my attention.

10 "American and German town planners," *TP&H*, 1(31 May), 1915, p. 19.

11 Sighted 1998 at the Powerhouse Museum, Sydney, see Watson (1998). The lithographed perspective drawing was published in *Building*, 15(12 November 1914), p. 63, and in a number of U.S. publications since 1893.

12 Freestone/Hanna (2008), p. 35.

13 Louis Sullivan, "What is Architecture?", *Building*, 14(12 June 1914), pp. 77-82; ibid, (12 August 1914), pp. 59-63; ibid, 15(12 November 1914), pp. 75-79; ibid, 15(12 December 1914), p. 60f; the articles are not recorded in Robert Twombly, ed., *Louis Sullivan The Public Papers* (Chicago: University of Chicago Press, 1988), or in idem, *Louis Sullivan His Life and Work* (Chicago: University of Chicago Press, 1986).

14 Sighted 1998 at the Powerhouse Museum, Sydney, with thanks to curator Anne Watson.

15 Taylor (1916), p. 189. Midway Gardens is best illustrated in Kruty (1998).

16 Taylor (1915), pp. 109-110.

17 George A. Taylor, "Some Chicago Curiosities," *Building*, 16(April 1915), p. 84.

18 Taylor (1915), pp. 109-110; cf. Taylor (1916), pp. 188-190.

19 Wright (1914); the first paper was Wright (1908), pp. 155-221, and an *ARecord* offprint.

20 Taylor (1915), p. 110; Wright (1914), p. 409.

21 George A. Taylor, "Eats and Eaters," *Building*, 15(12 July 1915), p. 63; also quoted Kruty (1998), p. 39, 209,n55.

22 Hanna (2002), p. 32.

23 George A. Taylor, "There," *Building*, 17(12 December 1915), p. 114.

24 As quoted in Taylor (1916), pp. 225-226. Larkin Building best illustrated in Jack Quinan, *FLW's Larkin Building: Myth and Fact* (Cambridge, Mass: MIT Press, 1987).

25 Taylor (1916), p. 227.

26 Byrne to FLW, 9 January [1914], p. 4, Wright Archive.

27 William Purcell, "Memorandum—insert 1913 or 1914," Purcell papers, as quoted in Freestone/Hanna (2008), pp. 145-146.

28 In 1915 Stowe gave a talk to the TPA of NSW about his "Visit to America with lantern slides", Freestone/Park (2009), p. 52.

29 On Marion see Anna Rubbo, "Marion Mahony: a larger than life presence," Watson (1998), pp. 40-55; idem, "Through the looking glass of 'Magic of America': Marion Mahony Griffin's role...", Turnbull/Navaretti (1998), pp. 338-346; Frederick Gutheim in Barbara Sicherman, et al, ed., *Notable American Women. The Modern Period* (Cambridge, Mass: Harvard University Press, 1980), v.4; here and there in Spathopoulos (2007).

30 The only surviving record of TPA meetings is secondary: "Women and Town Planning the sociological side", *TP&H*, 1(12 April 1915), pp. 28-30; see also Florence M. Taylor "A Message to Women," ibid, pp. 25-26; "Women in Town Planning," ibid, pp. 28-30; [F.M. Taylor], "Women in Town Planning," ibid, 1(31 May 1915), pp. 3-4.

31 "A tribute to a team," *TP&H*, 1(May 1915), p. 18.

32 Bui*lding*, 14(12 January 1915), p. 103, with thanks to Christopher Vernon.

33 Florence M. Taylor, "American kitchens versus Australian kitchens," *Building*, 16(February 1915), p. 83.

34 Freestone (2007), p. 62; Freestone (1995), p. 265, who also referred to a "Town Planning" article in *Australian Women's Weekly*, (12 June 1915), p. 15.

35 Freestone/Hanna (2008), p. 149, as recounted by Marion in Griffin (1949), part IIA, pp. 60-61.

36 Griffin (1949), v.2, p. 24 (with thanks to Anna Rubbo), or v.4, p. 125 in Turnbull/Navaretti (1998), p. 43. For a general outline of events see Freestone/Hanna (2008), pp. 147-150. The Taylor files in the Mitchell Library, Sydney, do not contain material prior to the 1920s and preserved papers of the Town Planning Association of N.S.W. begin 1919.

37 George A. Taylor, "The Fight for Canberra," *TP&H*, 1(26 July 1915), p. 35; idem, "Town Planning and common sense," ibid, 1(27 September 1915), pp. 19-20.

38 Freestone/Hanna (2008), p. 141.

39 "Canberra," *Building*, 16(12 June 1916), pp. 48-50; "Canberra Royal Commission," ibid, 17(12 October 1916), pp. 50-51; Reps (1997), chapter 3 and 6; Paul Reid, "Walter Burley Griffin's Struggles to Implement His Canberra Plan, 1912-1921," in Turnbull/Navaretti (1998), pp. 22-23; Donald L. Johnson, *Canberra and Walter Burley Griffin. A bibliography of 1876 to 1976* (Melbourne/New York: Oxford University Press, 1980), items 223-233.

Another attempt to discredit the Wright School of architecture was through the catalyst of one of W.G. Purcell's Minneapolis houses, see "The home, An Australian American comparison, what the [American] 'New School' means," *Building*, 19(12 January 1917), p. 73; cf. Johnson (2012).

40 Louis C. Mullgardt, "A Building That is Wrong," *Architect & Engineer* (San Francisco), no.17(November 1922), pp. 81-89, abridged reprint with editorial comment as "Buildings that are Wrong," *Building*, 32(April 1923), pp. 68-71.

41 "Freakish and Faulty Architecture," *Building*, 15(December 1915), p. 75; see also "Australian Architecture. Where it stands and falls II," ibid, 18(January 1916), p. 51; "Australian Architecture; Whose to Blame?," ibid, 20(March 1918), pp. 58-59; "Unsound Architecture. The bungle of the Melbourne Catholic college," ibid, 20(March 1918), pp. 60-69.

42 F.W. Fitzpatrick, letter condensed in Florence M. Taylor (editorial attribution and article reprinted without illustrations in Stephen/McNamara/Goad [2006], pp. 221-227), "Buildings that are wrong: Criticism of freak architecture," *Building*, 32(April 1923), p. 71, and letter content reviewed in *Washington State Architect* (Olympia) where editorially the hotel was described as "atrocious" and the "dream of a narcotic imagination," as

reported in "More Anent...," *Architect & Engineer* (San Francisco), 20(February 1923), pp. 83-84. See also editorial (attributed to Florence M. Taylor), "Freak Architecture: Its contempt for sentimental association and correct principles," *Building*, 38(October 1925), pp. 68-80, reprint without illustrations in Stephen/McNamara/Goad (2006), pp. 227-237; F.W. Fitzpatrick, "The thrall of the axis," *TP&H* 1(25 October 1915), pp. 25-28.

43 Hanna (2002), p. 52,n.41; Freestone (1995), 266; Roe (1984), pp. 193-194.

44 George A. Taylor, "The Common Ground," *Building*, 15(12 April 1915), pp. 48-49.

45 Taylor (1924), p. 42.

46 As quoted in Johnson (1977), pp. 114, 115; Navaretti (1998), pp. 156-158; Donald Leslie Johnson, "Some Early Buildings at Duntroon," *Royal Military College of Australia Journal* (Canberra), 1976 annual, pp. 81-83.

47 James Weirick, "Spirituality and symbolism in the work of the Griffins," in Watson (1998), p. 77.

48 Aitchison (1972), p. 41. And yet the Griffins were asked to design an American Pavilion for the State War Council for "Our Boys" Day, Melbourne, erected 15 March 1918, Navaretti (1998), p. 155.

49 As quoted in Aitchison (1972), p. 77.

50 As quoted in Freestone/Hanna (2008), p. 144; see Griffin (1949), part II, p. 22.

51 Ray and John Oldham, *George Temple-Poole* (Perth: University of Western Australia Press, 1980).

52 Wright did not personally reply, Isabel Roberts (secretary) to Purcell, 12 October 1909, Purcell papers.

53 W. Griffin to Purcell, 6 July 1919, p. 1, copy generously supplied by the late David Gebhard, also in Northwest. While not critical herein, Stowe's "written statement" may have been within a submission to the Royal Commission of Inquiry (held 1916-1917) into Canberra affairs that readily cleared Walter.

54 As quoted in Freestone/Hanna (2008), pp. 145-146.

55 "The One-Roomed House, is it desirable?", *TP&H*, 1(28 June 1915), p. 10, reply p. 11, not illustrated; also 16(14 August 1915), p. 103, reply p. 104, not illustrated. The *Herald* text is republished in Griffin (2008), p. 260. A brief mention of Stowe is in Robert Freestone, *Model Communities. The Garden City Movement in Australia* (Melbourne: Nelson, 1989), p. 112; "Cheap—and pleasant—housing," TP&H, 1(31 May 1915), pp. 28-29.

56 Apparently Stowe held some form of membership in the American Society of Engineers.

57 M. Griffin to Purcell, 7 August 194_ (probably 1947), p. 2, copy from the late David Gebhard. Marion returned permanently to the U.S. in 1939.

58 M. Griffin to Purcell, 14 August 1949, p. 1, Purcell papers. A few months after the Taylor's visit Purcell sold up and settled in Portland, Oregon, where he conducted a limited architectural practice.

59 Griffin (1949), part IV, p. 418.

60 Griffin (1949), part IV, p. 378.

61 Griffin (1949), part IV, p. 42.

62 Vernon (1998), pp. 10-11; Maldre/Kruty (1996), p. 31, 36n.45.

63 Wright (1908), p. 164, who probably referred to the years of Marion's full-time employment. See also photographic portrait ca.1898 (probably by Wright) of Marion and Mrs.Wright in Paul Kruty, "Chicago 1900. . .," in Watson (1998), p. 13.

Sadly, it is not uncommon for people to be prejudiced by Marion's singular personality and inability to write clearly. One of the most ill-mannered and irresponsible reviews to come to this author's attention said that Marion "Mahoney's" (sic) description of events was "mush," that her "unpublished statements" were "incoherent, vitriolic and selective," and further that to suggest Wright may have taken an idea from another person was "to fabricate history," that to dismiss the work of "Nelson borders on the irresponsible," etc.; copy of anonymous review to editor of *JSAH*, 13 February 1998. Coherent criticism is always welcome, yet this was accepted by *JSAH* as valuable.

64 Schindler to Neutra, 12 March 1921, as quoted in McCoy (1979), p. 132.

Relevant to overseas participation, some difficulties were posed by the R.I.B.A. in London over details in the competition program that were resolved well before the submission date, Reps (1997), pp. 76-84 and Appendix B. Of 137 entries mainly from engineers, landscape and building architects, overseas entrants included Eliel Saarinen (Finland, 2nd place), Alfred Agache and A.L. Bérard (France), William H. Gummer (U.K. later New Zealand gold medalist), James Gibson (South Africa), Christopher J. Yorath (U.K. later Canada), F. Edward Masey (U.K. later South Africa), Nils Ostt Gellerstedt (Sweden), and from the U.S. there was Arthur Comey, Harold van Buren Magonigle, Bernard Maybeck, Emil Mische, Herbert Kellaway, and one Edward Bellamy whose submission has been lost; see Reps (1997), passim. All were first class architects and planners except, perhaps, Bellamy: is he the one who looked backward?

65 For examples of the Griffins' city and community planning see the many entries recorded in Kruty (2003), Vernon (1998), Navaretti (1998), and Johnson (1977). Wright's city and community planning efforts are principle subjects of this author's continuing research, Johnson (2004); of the 1930s see Johnson (1990).

4

Tiles and Blocks

> **There was Natco's terra cotta *textured* tiles, *tex-tiles*, and *textile* blocks; and then the Griffin's concrete *knitlock*; then Lloyd's concrete *knit-block*; and then came Wright's and Lloyd's concrete *textile* blocks and *textile tiles*.**

Architects Walter and Marion Griffin were experienced in detailing the application of the relatively new, increasingly useful and popular construction material of reinforced concrete, either precast or poured *in situ*, that is a wet mix poured into wood formwork constructed at a building site. At the time *in situ* was called "monolithic", the preferred term. For many years the Griffins were occupied in Wright's office working on details or supervising the construction of concrete structures such as Unity Temple, E-Z Polish Factory (Chicago, 1905), two Universal Portland Cement Company exhibits in Buffalo and Madison Square Garden, and the Larkin Administration Building in Buffalo. Or they were similarly occupied in other architect's offices.

Beginning as an independent practitioner in March 1906 Walter accepted architectural commissions in America and, after 1912 with his architect wife Marion in Canada and Australia. Their building designs were mainly for houses and large or small commercial structures and many employed concrete or masonry blocks of concrete or of terra cotta. In 1911 Walter was invited to submit a proposal for an exhibit of fairly good size for the Universal Portland Cement Company. The unrealized design of June that year employed large three-dimensional forms covered with cement stucco and a floor of (presuambly) concrete pavers.[1] Those experiences together with expert advice obtained from engineering and construction collaborators during design and construction phases is accepted as obvious, not just for the Griffins but the Wrights.[2] Knitted to the historical and verbal record discussed earlier herein, there must be an examination of other masonry available commercially off-the-shelf or those under consideration before 1923.

KNITLOCK

Walter Griffin has stated that before Canberra he had engaged in the practical study of concrete and clay products, mainly of masonry. He mentioned a dry-cast synthetic stone of his making had been tested at an unnamed university; tests of chemical admixtures for specific purposes; a study of cellular slabs; "partitions and dove-tail keyed hollow block walls, 'Natco' System" in 1906 and, he said, other ideas for "burnt clay."[3] More will be said of Natco. It can be understood, therefore, that like most of professional colleagues the Griffins were knowledgeable in basic construction technology. To venture further into invention would not surprise.

Design of the Griffins' concrete knitlock system began in late 1916 in Melbourne, Victoria, in collaboration with builder David C. Jenkins who helped with practical aspects and independently constructed a few small buildings using knitlock. The system was patented in Australia 31 May 1917 and a roofing concrete tile system on 2 March 1918.[4] Knitlock machinery used sturdy metal molds and an integrated press. Walter referred to the wall system as "Knitlock or 'Segmental Architecture.'" The flat roof tiles were technically and aesthetically unrelated to the knitlock wall system.

According to Griffin historian Peter Navaretti the first knitlock buildings were a pair of cleverly organized one room cottages for the Griffins as rental or sale units, traditional bungalow in appearance and constructed in 1919 near Frankston south of Melbourne.[5] That same year Jenkins built a house in the Melbourne suburb of Malvern. That September knitlock was demonstrated at the Royal Melbourne Show of Agriculture and Industry. The Griffins' own one-room cottage was begun probably in 1920 and completed by their hands early in 1922 on a site in the hills east of Melbourne at Heidelberg (see Fig. 4.1). Those cottages were followed by a number of unrealized projects and at least twenty-one buildings constructed in and around Sydney and Melbourne before Walter departed for Lucknow, India, in 1935, not to return.[6] The largest and most complete knitlock house that also used the system's roof tiles was designed in 1922 for Stanley Salter who occupied it in 1923.

Within the realm of masonry construction, knitlock was a significant invention, unlike any other in Europe, Australia or North America. That was why patents were granted. It was a structural system not of masonry in compression but composed of a series of precast concrete tiles stacked vertically and horizontally and interlocked by means of their own shape and in straight coursing. In one piece of promotional literature of 1922 a knitlock house was described as having . . .

> *strong double walls throughout, internal and external, lap-joined and insulated, weather-proof, hollow, with air-circulation space; cool, silent, sanitary, with all corners coved. They are durable, fireproof, white-ant resistant, and resilient to withstand shocks.*[7]

Typical of promotional literature it lacked detail. After seeing a few of Wright's textile block buildings in Los Angeles in 1925, perhaps to clarify important

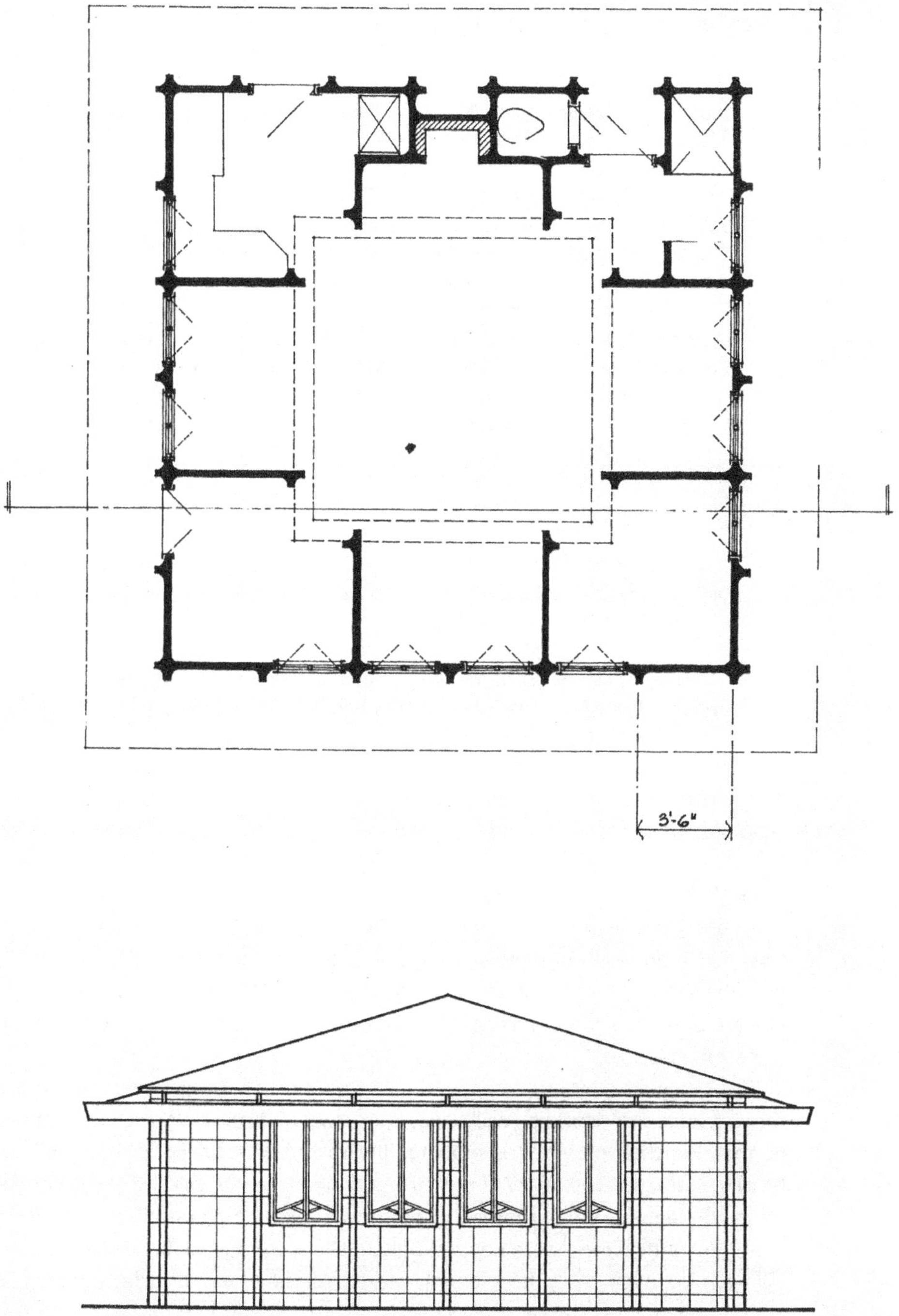

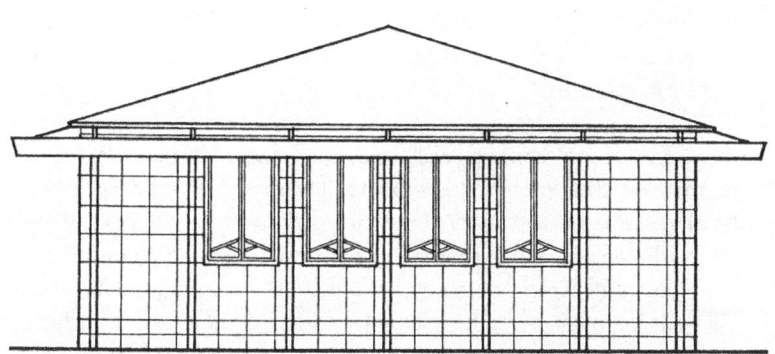

Fig. 4.1 Plan and elevation of their own one room cottage by architects Walter and Marion Griffin, 1920-1921

differences to his knitlock, Walter had a more useful piece published in 1927 in which he described his "Segmental Architecture" system as follows:

> *Technically there are only two types of segments—vertegral, which lock together to make the frame work or skeleton, and tesseral, which lock together for two-ply curtain walls, attaching to and stretching between the vertegral columns.*
>
> *The double wall provides for lap joints everywhere, and between the inner and outer layers there is an insulation layer of bitumen, as well as a proportion of air ducts in the interlocking keys, which allow for cooling by convection where the climate requires it. In these flutes also are concealed gas and electric conductors. In erection, the segments are simply slipped together from above, the vertical joints being filled by pouring in concrete grout, in special cases with steel reinforcing rods. With the manufactured half-segments, all the openings and corners work out accurately, so that, without cutting or fitting, the grooved frames of the doorways and windows and fixtures are likewise slid into place and made fast with grouting also . . . .*
>
> *Internal wall faces consist of six-inch squares . . . .*
>
> *Externally, the segments appear mainly as twelve-inch chequers with tuck-pointed joints, the effect . . . being cut stone, the colors being supplied by the sands selected for the surface . . . .*
>
> *The house of Mr. Salter, at Toorak . . . also demonstrates how . . . it was feasible to provide all the rooms with at least two external walls, and secure the sheltered, cool, cloistered central court garden . . . .*[8]

Bitumen would also help reduce the transfer of moisture by sweating or capillary action from the exterior. Joints between blocks were treated with ordinary mortar. For comparison Griffin illustrated his article with a photograph he had taken of Wright's Freeman house of 1923.

Patent drawings 12-17 in Fig. 4.2 are "vertegral" members, drawings 1-10 "tesseral" members. All drawings are plans (that is cut horizontally and looking down) except 2, 3 and 9. When in place the twelve inch wall tiles lapped by fifty percent horizontally and vertically. So the three-dimensional construction module was six inches.

The total width of a wall, that is the tesseral two-ply components that formed the non-bearing "curtain wall," was but two-and-one-half inches. Although nonbearing they provided lateral stability to the vertegral or column-like load bearing elements. The cavity of the vertegral members was packed with a concrete mix and steel reinforcing rods and performed like structural pilasters to inhibit bending. Vertical rods were needed only in the tesseral wall "in special cases," for tall walls for instance, or those mainly glazed. All the exposed surfaces were plain, without colored aggregate (only sand as binder, filler and texture), without relief, unadorned. The construction process was very difficult as it was for Wright's later versions of textile blocks. Tolerance was not available in plan dimension or vertically at ceiling or joist plate line, that is the constructional three dimensional X-Y-Z axes.

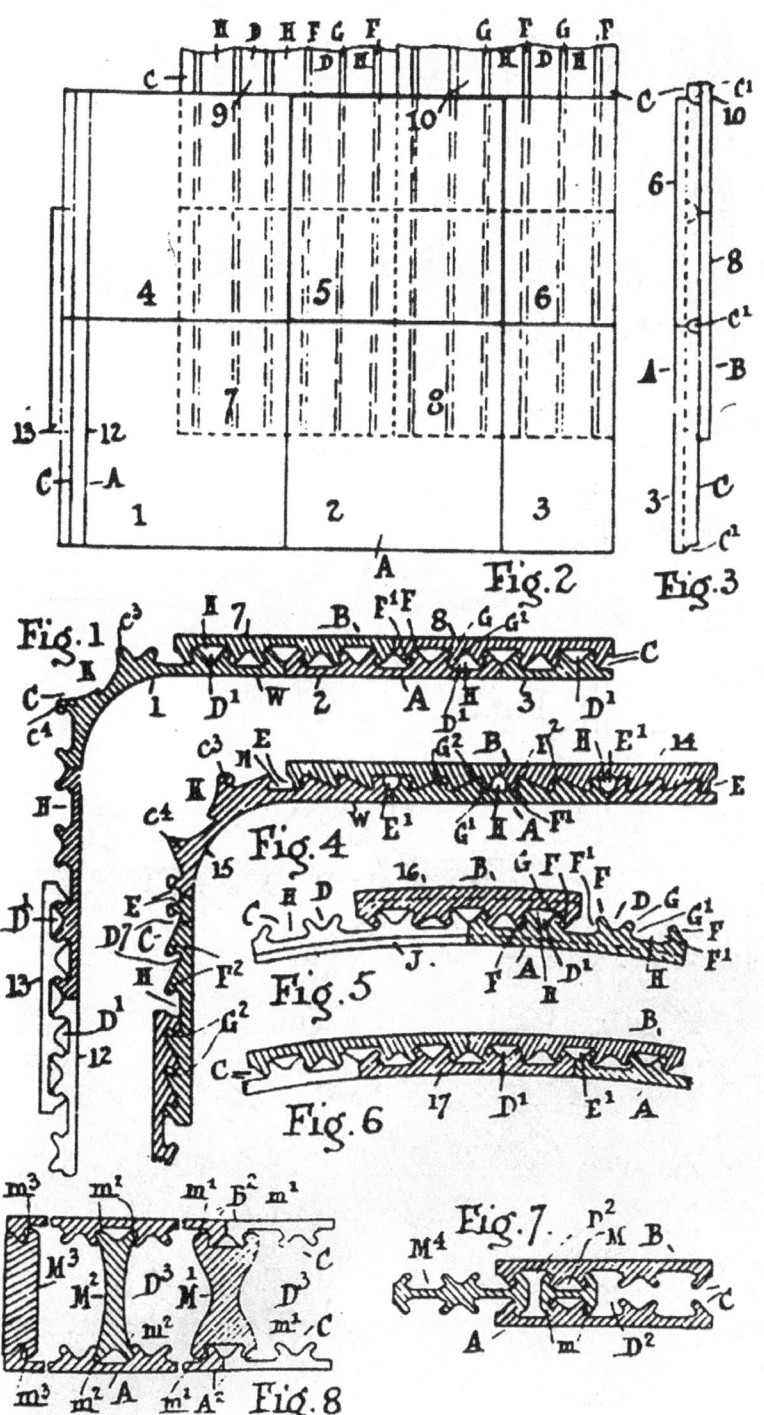

Fig. 4.2a Knitlock construction system, patent drawings as published in 1917

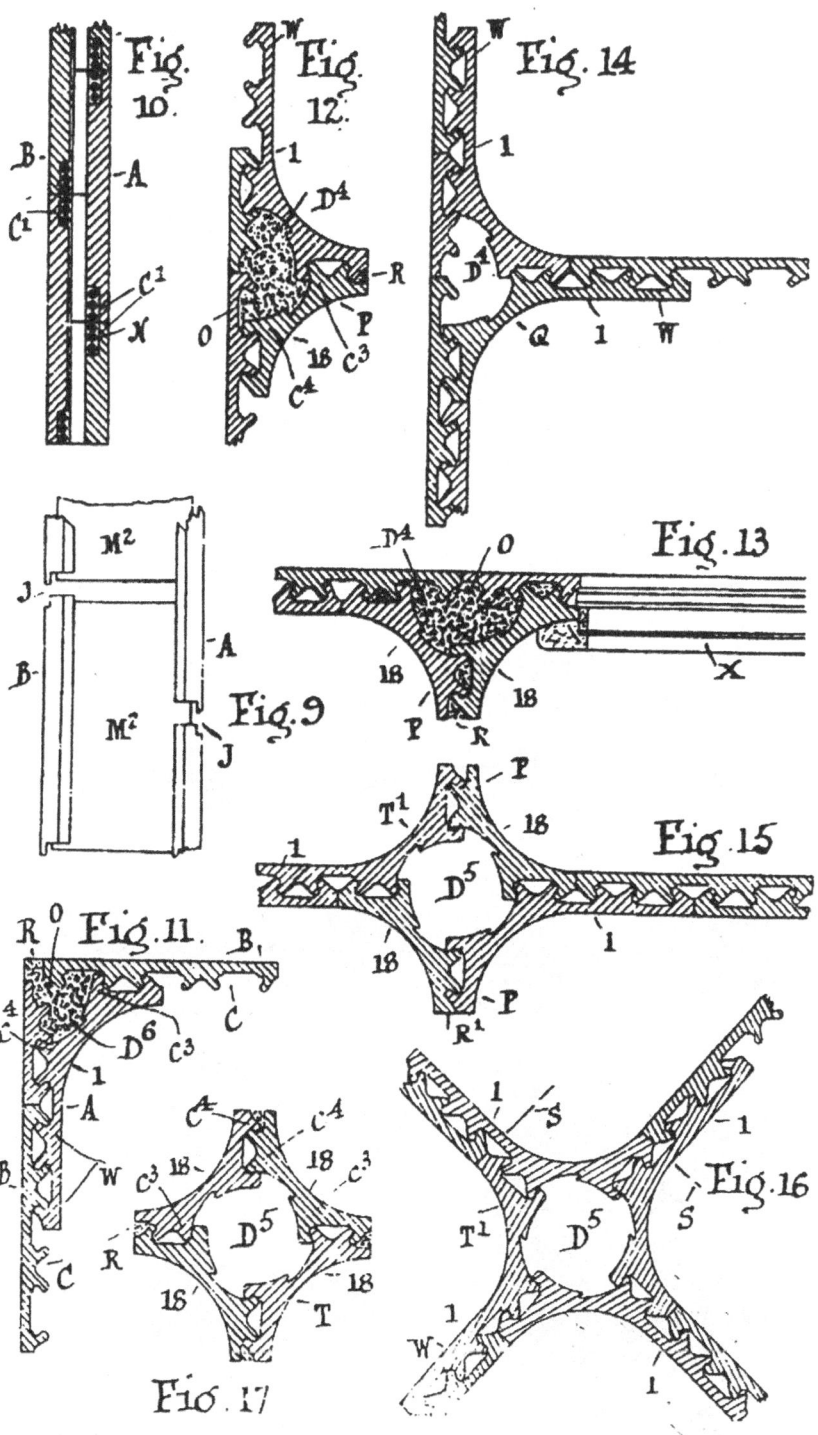

Fig. 4.2b Knitlock construction system, patent drawings as published in 1917

Confusion now and then arises in the use of the terms "blocks" and "tiles". Historically tiles were made of split slate or split wood or fired clay shingles. All types were thin, small in size and applied (since antiquity) to walls or, when larger, thicker and overlapped, for roofing. The term "blocks" traditionally referred to rectilinear stones used constructionally in compression for foundations in the earth and above in the building proper as masonry for walls or in arches. Like hollow clay masonry and tiles, after 1900 concrete blocks were used structurally and non-structurally. Generally blocks are self-supporting, tiles are not. But in practice there was little distinction between the terms "tile" and "block." By around 1900, when discussing walls they were interchangeable, at least in America.

Because they were narrow and *individually* not self-supporting laterally or collectively very weak in resistance to bending, each tesseral knitlock piece was a tile that had to be locked to facing neighbors. Son Lloyd's knit-block was a block and Wright Senior's initial textile blocks (for the Millard house) were blocks and they did not lock together horizontally. Senior's later designs (up to the date of Florida Southern College and the Levin house) were thinnish and more accurately described as stackable tiles. To repeat: a single withe of tiles can not be load bearing. For the Millard house his textile masonry was used as both compressive (self-supportive) and as tiles, some glued to lintels. Regardless, the Griffins and the Wrights used both terms, block and tile, without consistency.

NATCO

While the Griffins patiently struggled in Australia, certain events in America were unfolding that would influence Wright. But those events, as historian Miles Lewis has shown, were not associated with concrete blocks but with fired terra cotta, i.e. clay.

The National Fireproofing Company of Chicago, or as they condensed, Natco, published *The Natco tex-tile one-family house* in 1917. The book was "a selection of designs submitted in competition by architects" together with "houses built of Natco tex-tile" in the United States. Interestingly, one of the competition entries was a competent Wright-School design submitted from Melbourne by George Elgh, a Chicagoan who in 1914 had traveled with the Griffins to Australia where he worked as a draftsman in their private Melbourne office.[9]

Also, the name "tex-tiles" was given to a new line of Natco hollow clay tiles. The term "tex-tiles" was derived from Natco's earlier "Textured Tile" product. "Hollow tile" wall blocks, they were a modest redesign of their standard hollow clay masonry. They had hollow chambers and trapezoid-shaped surface grooves on their interior and exterior faces to better hold a cover of cement stucco or plaster. Walter Griffin referred to them as "dove-tailed grooves" and, as noted, apparently they were his contribution to Natco. Wright's Barnsdall house of 1919-1922 is a good example among tens of thousands that used stucco troweled onto Natco tiles. Clay tex-tiles were also manufactured in a style to compete with the growing popularity of concrete blocks. They were meant to eliminate the plastering trade by giving the tex-tiles a rough finished exterior face that was exposed to look similar to certain bricks and concrete blocks. The interior surface was dove-tail grooved to receive plaster.

Then in 1919 Chicagoan Charles E. White, Jr., an architect who had worked alongside Marion and Walter in Wright's office 1902-1904, wrote for the clay tile industry a book on *Hollow tile construction*. It included two photographic illustrations of Walter Griffin's Rule house in Mason City, Iowa, and one of his Gunn house at Beverly, Chicago, both of 1910. The walls of each were constructed of terra cotta, i.e. hollow clay tiles with dove-tail grooves, not tex-tiles, and then stuccoed. White also discussed and illustrated H-shaped tiles that had a recessed edge groove on two sides to receive reinforcing bars if necessary. Also illustrated were curved "silo blocks" that also had an edge groove to receive steel flat bars as reinforcing. Those bars ensured walls would hold shape against internal pressures exerted by a fully filled silo cylinder. White also presented "textile blocks," but note Natco's change in spelling to "textile" and the word "blocks."[10] All this confirms that the Griffins and Wrights were familiar with Natco's products, with edge grooves to receive reinforcing, and other clay construction products: the terms "clay" and "terra cotta" were then used interchangeably.

Wright understood the connection. "Cast terra cotta" and "concrete," he said in 1928, "the two materials have much in common. Terra cotta having the great advantage of standing up to be modeled and becoming indestructible [?], colorful and glazed when fired . . . ." One must assume that "indestructible" referred to the material when applied as fire proofing steel and that "colorful and glazed" referred to its application as architectural decoration. Anyway, he reckoned his concrete textile houses "stand delicately perforated like a Persian faience screen," i.e. walls like a Persian screen that in tradition were made of fire-glazed terra cotta earthenware.[11]

KNIT-BLOCK

According to curator Robert Sweeney, Wright's initial textile block was designed for the Millard house during early 1923.[12] We have outlined how it was preceded by hundreds of precast clay and concrete designs and the Griffins' singular effort. But equally influential was Wright's son Lloyd who during 1922 designed a bulky, chunky, inelegant house for Henry Bollman in Hollywood.[13] Bollman was a builder and preliminary construction drawings dated 1 December 1922 show lower walls of "stone blocks." (Lloyd called himself a "designer" because he was not yet a licensed architect.) Those lower walls Lloyd described as a "knit-block [construction] system composed of thin 4" blocks with an air space between and tied together by steel."[14] This suggests that "knit-block" was probably something like a typical masonry cavity wall construction. Yet Lloyd has also stated that his knit-block masonry was "in cast block similar to the [later] Millard house."[15]

Knit-block's exterior surface was relatively uneven, without a pattern in relief, plain and soft-edged with no exposed aggregate. It intentionally appeared somewhat like adobes of sun-dried mud. One commercial product was called "Adobe Stone". It appeared soft-edged like adobe and sometimes referred to as "slump block". Structural and architectural references to Southwest Indian adobe architecture are also evident in Lloyd's Oasis Hotel in Palm Springs, California, designed in 1922 and finished in 1923. A slip-form method of concrete construction was employed at Oasis in such a way as to suggest linearly stacked adobe. During his developmental years Lloyd had been involved with Irving Gill on the construction of aesthetically plain, simple buildings some of which employed lift-slab concrete walls or stuccoed hollow clay blocks, notably for the new industrial town of Torrance, California. The town's master plan was prepared by another of Lloyd's former employers, the landscape architectural firm of Olmsted and Olmsted, that is Frederick Law, Jr. and brother John Charles.

Lloyd mentioned with certainty that after his father returned from Japan in August 1922, the Bollman house "inspired" him to pursue a "knit-block system." "[F]ather saw it," Lloyd said, "and saw that this concept could be worked into a total system...."[16] As the late historian David Gebhard correctly observed, development of textile blocks for the Los Angeles houses from 1923 to 1925 was a collaborative effort of father and son.[17] In 1928 Lloyd reported that he had "built eleven buildings with the blocks," presumably houses. He probably included Bollman and father's first four that were executed more-or-less cooperatively. For each of his own buildings, however, Lloyd mixed his knit-blocks with standard wood stud wall construction.[18]

TEXTILE BLOCKS AND TILES

The block system Wright designed for the Millard house in 1923 was borrowed from the Griffins. As well, and like Lloyd, Wright also adapted Walter's term. Since he too was working with masonry blocks Wright then borrowed Natco's term "textile". Further, Wright often used words such as "knit" and "weave" when discussing in confused language his "textile" blocks or modular design in general. In those discussions he imitated theoretical ideas that had been composed by many mainly European architects before him. All the terminological analogies revolved around loom-built cloth fabric textiles with their web of warp and weft, something Wright would soon exploit inaccurately in verbal descriptions.

The obvious conclusion from all this is that during 1919 the Taylors' friend Stowe had visited Wright's office with information about Griffin's knitlock. Schindler and Lloyd were present while Wright Senior was absent in Tokyo. In undertaking the design of precast concrete blocks on return from Japan, Senior "might have been spurred by some sense of rivalry" with Griffin, historian Miles Lewis has speculated,[19] but more likely it was with Lloyd. The same motivation cannot be said of Lloyd. A close friend of the Griffins since childhood, he was curious about knitlock's potential and wanted to conduct his own experiments. But in 1923 he and father had to be careful. Knitlock was a true invention, one so unique that anything too similar would have been obvious. Senior came close but quickly changed direction.

Fig. 4.3 is a reproduction of a blue print of Wright's patent drawing presumably prepared in mid-1923, certainly not earlier. Drawing 6 therein is a cutaway plan

Fig. 4.3 Blueprint of a textile block patent drawing of 1923, "Inventor Frank Lloyd Wright"

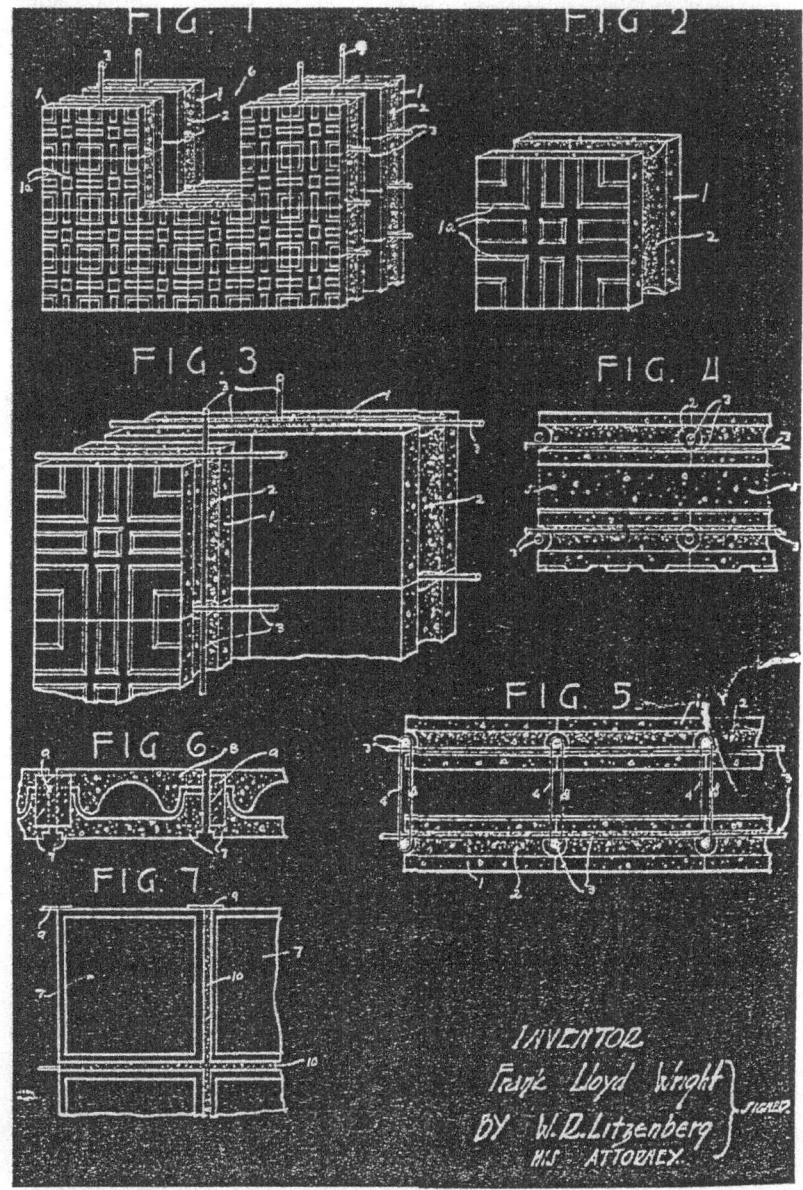

view through two contiguous tiles bound at their corners by a bit of wire mesh, i.e. expended steel plate, and set in mortar. All together the blocks would form a wall eight inches thick. This was used for most bearing walls of Millard's house. A comparison of Griffin's design with drawing 6 and photographs of the Millard, Storer and Ennis houses under construction shows the obvious connection. Many of the walls for those houses used only an exterior faced tile without its interior mate.[20] A practical problem, yes, but also a major theoretical problem to which we shall return.

Drawings 1 to 5 in Fig. 4.3 explain interior and exterior rows of blocks with an air or grout-filled cavity between, and they relate to C in Fig. 4.4. Drawing 2 is a single patterned-faced block, drawing 4 a plan section with concrete infill, and drawing 5 a plan section with tie wires across a cavity. Drawings 6 and 7 are of the first blocks used for Millard and relate to B in 4.4. Therefore, the set of drawings were, confusingly, for two systems, a no-no when applying for a patent. One system comprised drawings 1-5 (as used for the Freeman house only), while 6 and 7 were as used only for Millard. A preliminary design drawing of early 1923 for Millard, Fig. 2.2, contains the principle floor plan and the main and a side elevations. As well, at left are sketchy pictures and sectional details entitled "textile block construction" show the position of small squares of wire mesh at the horizontal joint (see left vertical joint in drawing 6 of b) and a perspective study of the blocks at a corner. Those details were transferred to the patent drawing.

Wright occasionally argued that patents were anathema in support of profiteers. While he did not make application to patent textile blocks and tiles, in the late 1930s he was granted four patents, each related to the Johnson Wax building, each of 1937 and granted 1939. And a design patent for a fourplex dwelling unit of 1938 was granted. No other patents are listed for Wright.

Griffin's patented knitlock employed vertegral members at corners for structural rigidity and continuity.[21] Wright's first Millard design, Fig. 4.4B, indicates lateral (horizontal) resistance was achieved by the shape of the male block fitting rather loosely inside the female, almost exactly like Griffin's. This fact probably saved a test of copying. Blocks after Millard did not interlock horizontally or overlap, were thin and best called tiles. In comparison to knitlock, continuity and lateral stability were obtained only by small discontinuous steel mesh at horizontal joints for Millard and afterward slightly pact steel rods to resist ground waves generated by California's frequent tremors, occasional earthquakes or land slides.

A modification of the Millard blocks was proposed during April and May 1923 when designing another Barnsdall house to have been in Beverly Hills,[22] redrawn here as D in Fig. 4.4. Dashed lines indicate blocks on either side similar to C, the implication is of a slight transitional design to E.

To a fellow architect Wright described without flourish how blocks were made:

> *The material I have used for the molded blocks in the Millard Home and others is a simple mixture of Portland Cement Concrete—one of cement to four of sand.*
>
> *The sand is rather coarse, sharp and clean. We like to vary the sand so the blocks will not be exactly alike. I use what is called the dry-mix that is just wet enough to take the shape of the hand and kept it when squeezed and this mixture is tamped into the molds by hand. It does not make a water-proof block used in this way, so we coat the inside of the block with asphalt when it is dry, and before setting the block in the wall.*
>
> *The texture of the wet mixture is not so agreeable a surface but it would be more waterproof.*[23]

There is no mention of aggregates from the site and he was correct to anticipate future problems.

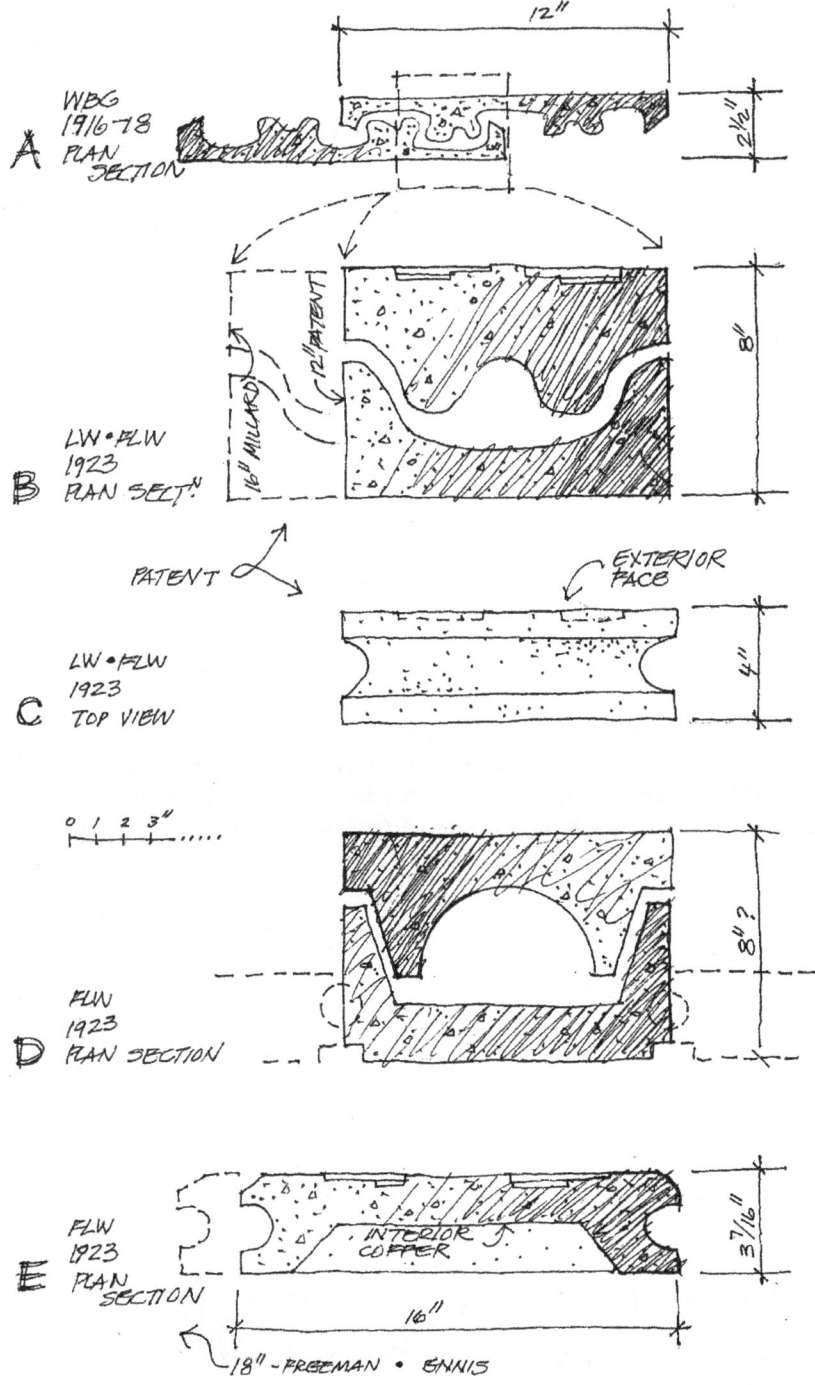

Fig. 4.4 Griffin's knitlock tiles (A) compared with Wright's concrete blocks and tiles, 1916 to 1924

Wright's enthusiasm for an organizational system and aesthetic proportion derived from the geometry of square and cube, leading to a module, has been presented and evaluated elsewhere.[24] For our purposes Wright's exploration began earnestly—or at least more obviously—in the 1890s with the design for small square molded glass blocks for the Luxfer Prism Company in Chicago.[25] The square, and proportions derived from it, held sway in his design imagination (at least until the 1930s) for decorative elements such as lighting fixtures, carpets, wall tiles in clay or concrete, window leading, and most other things receiving a pattern. An immediate predecessor to the textile block houses was the Imperial Hotel, Tokyo, where complicated relief designs for square wall tiles and blocks were cut in stone or precast in concrete or in clay. Some were suggestive of the later and more modest textile block designs. One tile for the hotel was cast in clay, perforated and used alternately with carved stone blocks as facing on piers. Those remarkable textural characteristics visually resemble the piers of the Storer and Ennis houses.[26] Like other artists Wright's designs were continually evolving even when reexamining, redesigning, and reemploying.

From their beginning in 1923 each of the Los Angeles houses had block face patterns that were cast in relief.[27] Three different square faces were designed for the Storer house, Fig. 4.5. One (C) was reused by Lloyd (together with wood stud construction and the original Millard block designs) for an ample display room addition to Millard's house and built during 1925 and 1926.[28]

Prior to exploring concrete blocks, Wright studied native, indigenous and exotic idioms to enrich his understanding of the essentials of architecture and two dimensional design, a theme to which we will return. A related search focused on his mother's Welsh roots and the early ornamental art of Wales. (He completely ignored the English and New England roots of his father.) A modest example of those studies is found in his initial designs for the textile blocks' face.

Derivations of the carved symbol found on the birth stone of the illusive sixth century Welsh poet, Taliesin, was widely used in Christianized Celtic art.[29] As well, Medieval Norwegian artists applied it in all decorative media; Viollet-le-Duc used it as a decorative motif; and it was found in popular Western designs derived from oriental art, oddly enough, and British Arts and Crafts influences. It was the basis for Wright's personal logo first designed in the 1890s, and for tile faces for Millard and Storer. But the cross was imaginatively joined by another design called "pure." Art historian Marie Frank effusively described pure design as

> *a formalist pedagogical method, relied on exercises with abstract design elements—the dot, the line, shape, and color—to encourage the creative ability of students. It depended on, or grew out of, contemporary investigations in science, psychology, art education, and philosophy. Considered within this congeries of activity, Pure Design represents an important shift in American aesthetic thought affected how architects regarded architecture.[30]*

Begun as a discussion group in 1885 and continuing as an informal school around 1895, in 1898 the Chicago Architectural Club began operating ateliers conducted by architects Robert Spencer, Dwight Perkins and Mat Dunning. The club was keen

Fig. 4.5 Drawing of four relief patterns on exterior faces of Wright's blocks and tiles

on political arguments put by the Arts and Crafts Movement and resulting designs. In 1904 the club opened a suite of meeting rooms downtown and attracted some of the most progressive of young draftsmen. In contrast to the professional associations of older practitioners, the club's "concern was design" in which would be found "the solution of social or functional problems by architectural form."[31] The desire for that solution cannot be over emphasized. The mode of search was not found in traditional schools of architecture or the hidebound American Institute of Architects, but in self education. Members included not only the young atelier masters but counterparts such as Myron Hunt, Webster Tomlinson, Walter Griffin, Allen B. and Irving K. Pond, Thomas Tallmadge, a converted Peter B. Wight, Dwight Perkins, George Dean, and Wright. It was around 1899 the club "proclaim[ed] a new principle that might revolutionize architectural design," a principle known as "pure design."[32]

Parallel to and supportive of the club, landscape architect William French had begun an architecture program at the Art Institute in 1898, initially directed by Sullivan's one-time collaborator Louis J. Millet, a former student at the Paris *École des beau arts* and a close friend of Viollet-le-Duc. Then in 1899 French hired Emil Lorch, Massachusetts Institute of Technology graduate who also studied at the *École*, as assistant director and to manage the architectural program. With other young men Lorch mounted exhibitions beginning 1900 in the Arts and Crafts room and separately a display of Wright's work. In 1901 their exhibition catalog included a greatly expanded version of Wright's essay "The Art and Craft of the Machine"; his buildings were again separately mounted in 1902.

Early in 1901 French and Lorch encouraged the Institute to sponsor lectures by Bostonian artist and educator Arthur Wesley Dow on two-dimensional composition in Japanese art. Dow offered "universal" two-dimensional compositions, often interpreting Japanese art, usually as found in inexpensive wood-block prints. He often used the geometry of the square as a basic shape and within presented possible spatial or linear configurations. Millet then modified the Institute's architectural program that was based on the Paris École system of beaux arts to one that incorporated Lorch's understanding of pure design.[33] Student and historian of pure design and Dow, Kevin Nute has said:

> Essentially pure design entailed a reduced emphasis on historical styles in favour of a neutral vocabulary of simple geometric forms, and significantly Dow himself was directly associated with this movement.... [He] was invited to address the third annual convention of the Architectural League of America held in Philadelphia in May 1901, a conference which was dominated by the topic of pure design.[34]

Robert Spencer and Emil Lorch, two of Wright's colleagues in the Chicago Architectural Club, presented papers. Dow accepted the universal principles espoused by the expert on Japanese art and preservation, Ernest Fenollosa, saying that,

> [Fenollosa] believed Music to be, in a sense, the key to the other fine arts, since its essence is pure beauty; that space-art [painting] may be called "visual music," and may be criticised and studied from this point of view. Following this new conception, he had constructed an art-educational system radically different from those whose corner-stone is Realism. Its leading thought is the expression of Beauty, not Representation.[35]

Wright was an ardent student of Japanese traditional art and a collector of wood-block prints.

In March 1901 the club, now housed within the Art Institute's building, sponsored an afternoon's program on pure design and the idea of "progress before precedent." Dow gave a lecture and exhibited some of his work.[36] A seminar took up the question "how can pure design be best studied." Answers were as varied as the men who spoke. Their understanding not helped by one of the founders, so to speak, of Pure Design, Denman W. Ross, who said: "By Design I mean Order in human feeling and thought" and how they are "expressed." And again: "By Pure Design I mean simply Order, that is to say, Harmony, Balance and Rhythm."[37] If simply Order, why use the term pure design? Like Dow, Ross linked order to music: Pure Design appeals to the eye just as absolute Music appeals to the ear." But he did not define "absolute".[38]

Lorch took the discussion from just two-dimensional design and to architecture when he put it so:

> In architectural design instruction—how to strengthen and draw out the analytical power, the appreciation and invention of the architectural student...

> *how to make him not an <u>adaptive</u> but a <u>creative</u> worker [he emphasized], or an artist-builder—an architect.*[39]

From the techniques found in Dow's system of two-dimensional lines, patterns and visual effects, Lorch went on to three-dimensional exercises in the molding of form: cylinder (column), parallelepiped (rectangle), cube (square) and other geometric and natural and rhythmic conventions that might freely combine into complex compositions. From the record of his work we know this intrigued Wright. His elliptical post-mortem rationalization was that

> *Design is abstraction of nature-elements [in] purely geometric terms—that is what we ought to call "pure design" . . . But—nature-pattern and nature-texture in materials themselves often approach conventionalization, or the abstract. . . .*[40]

He probably found the debates and seminars interesting but of limited stuff, not logically transferable in a useful way to architecture. Lorch's practical extension was another matter. In practice pure design principles are found and exemplified in many obvious ways in nearly all of Wright's work post-1900 and that includes the four houses in Los Angeles.

When it came to action Wright was thoroughly practical. He simply borrowed and remodeled two-dimensional designs from Dow's book of exercises and gave them three-dimensional vitality. One page from Dow's book on *Composition* makes the point.[41] Fig. 4.6. The two-dimensional design of all concrete blocks began with a basic geometrical design taken from Dow; so too the designs for wall tiles at Midway Gardens in 1913; so too some tile designs for the Tokyo Hotel. And where else?

The Storer face design (Fig. 4.5A) was used earlier as precast concrete tiles that cap the exterior wall of the 1915 A.D. German Warehouse in Richland Center, Wisconsin, and wall tiles at the Tokyo hotel. They in turn were derivative not only of Dow's studies in pure design but less effectively of varied sources in Mediterranean and pre-Columbian antiquity. The Storer design (Fig. 4.5C) is an adaptation of elements found on the upper part of the shaft near the capital of the window colonettes on Unity Temple where flora was conventionalized.[42] Fig. 4.5B is a more elaborate asymmetrical design not directly related to Dow and used on the Freeman house.[43]

Even though it is a rather small building, the Freeman house required 12,000 handmade, sixteen inch solid tiles (not coffered) of more than fifty-six different shapes with faces plain or in relief or glazed. That total is twice the number if the walls were made of the usual six or eight inch single block width. Also, there were eight times more individual shapes than for normal block construction. We should note that patterned blocks were used in conjunction with plain-faced blocks on the walls of all four houses.

Fig. 4.6 Two-dimensional "Line-ideas" composed within a square by Arthur Wesley Dow.

APPLICATION

However, serious displeasure with the construction process, structural concerns, extremely high costs for precasting blocks (this was forcefully put by each client) and for wages, were among factors that impelled the Wrights to redesign their system. As we shall learn, their revisions only compounded most problems. Wright confessed these problems in 1927 when referring to the four houses as "experimental" models. In a letter to Chicago architect A.N. Rebori who was preparing for *Architectural Record* an article on concrete block houses, Wright moaned that; "We had no organization—Prepared the moulds experimentally." The work "was roughly done and wasteful," and he confessed "The mixture was not rich—Nor was it possible to cure the blocks in sufficient moisture," and there was "trouble" in "making the buildings water-proof." And so forth, see Appendix 2.

For the Millard house a crude mold was made on site of wood pieces nailed. The later houses and the unbuilt playhouse (and soon the Phoenix hotel) employed pressed metal molds. We've noted that the mixture was too dry and that Lloyd insisted on watering the drying blocks and tiles as they cured. Wright also informed Rebori that he and Lloyd had put too much of their time and effort "on invention".[44]

More importantly, there was a lack of organization on and off site, difficulty installing the blocks, a lack of quality control and poor workmanship: all this Wright laid at the feet the person who supervised construction. However, Wright was in charge of all aspects of the design and construction processes and that included the manufacture and installation of molds and blocks. Wright briefly supervised the Millard house but quickly delegated the task to Lloyd. At all stages of construction Lloyd kept father informed by mail and telegram and sought advice when necessary. Wright Senior was in attendance in Los Angeles only occasionally from 1922 in to 1924 when he decided to return to Wisconsin, leaving Lloyd isolated but probably happier. So, during the years under discussion here, and as their correspondence reveals, Lloyd was the fall guy.

At the time of writing to Rebori Wright had just been hired as a consultant on the design and detailing of a resort hotel in Phoenix, Arizona, the Biltmore, to be constructed using his textile block system. He was worried, perhaps, about possible litigation for plagiarism. He was getting a divorce and planning a second marriage. Out of work, he was scurrying about trying to talk people with money into financing some personal construction proposals, all to no avail. He was hardly in a positive frame of mind.

A drawing detailing aspects of textile tile construction was prepared in 1925 for publication in Germany by Austrian emigrant architect Richard Neutra, who had begun in Wright's Chicago office in October 1924, Fig. 4.7. Three versions are known: one in metric measure and German text; one metric with English text; and one all English. Each explains the revised textile block system as a cavity wall composed of lighter coffered tiles, here showing the Ennis relief pattern.[45] The revised tiles were first employed on the Storer house. It was begun in late 1923 and developed simultaneously with the partly constructed kindergarten.[46] The system Neutra illustrated was the basis of the design and installation of the Arizona Biltmore Hotel, of Lloyd's variations of 1931, of modifications for Florida Southern College, and for the Levin house of 1948. While intriguing, those post 1924 applications are well outside the parameters of this case study.[47]

By 1924 the "block-slab" system was only slightly more practically resolved but certainly not economically in energy, time or dollars. Even in the late 1920s Wright remained optimistic, at least in print, as he explained: small molds with "any scheme of pattern or texture imagination conceives" can be made of a weight a worker can carry, like a clay brick. (But terra cotta is fine grained, easily worked when damp, superior for casting in molds, weighs less, but when fired lack tensile and shear strength.) Further, as Wright explained,

> Grooves are provided in the edges ... so a lacing of continuous steel rods may be laid ... for tensile strength. The grooves ... may be poured [actually hand compacted] full of concrete after each course of blocks is set up, girding and locking the whole into one firm slab.[48]

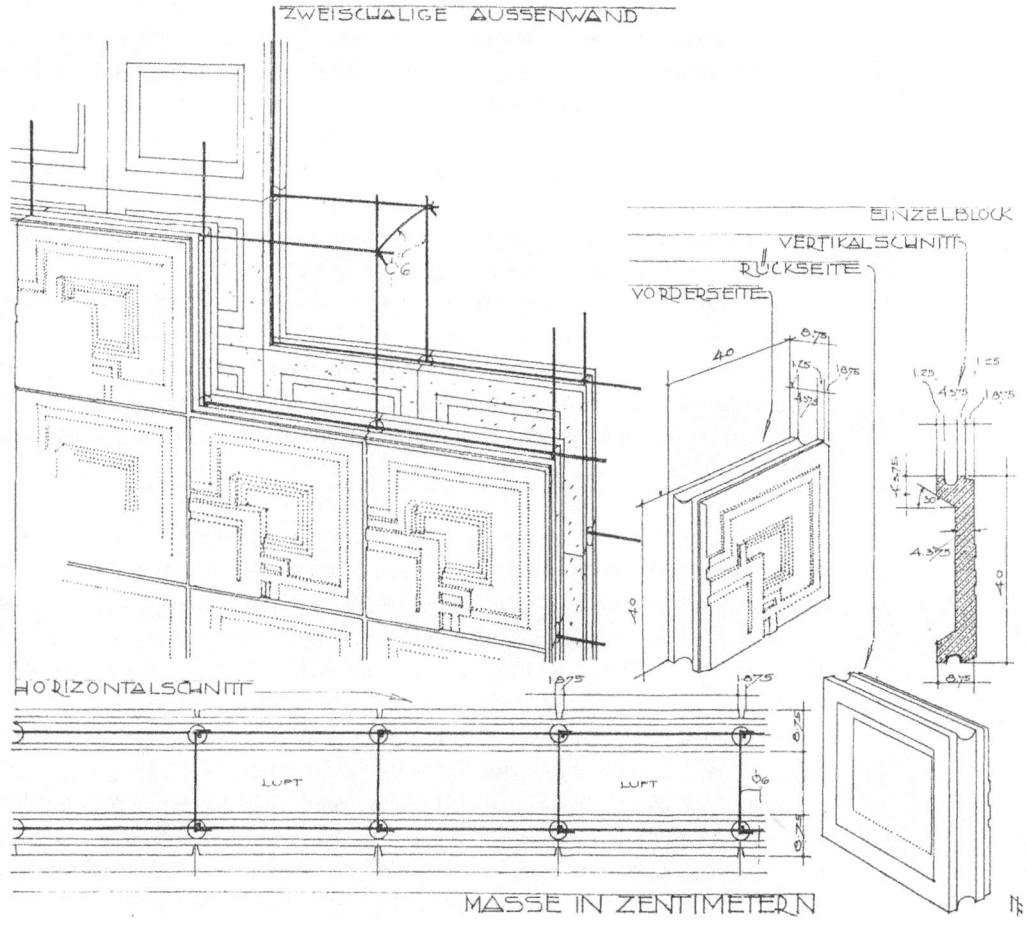

Fig. 4.7 Wright's wall tile system as drawn by Richard Neutra in 1925

Wright's use of Griffin's word "locking" is interesting. Wright's related idea, we note, was that the embedded two-way reinforcing rods created a "firm slab," a "textile-block slab construction." But its structural capacity as a wall was suitable only in the vertical. It was ineffective horizontally or in flexure, therefore not firm.

Also, consider this. Walls for Storer, Ennis and the proposed playhouse varied in width from sixteen to twelve inches (both with a cavity), to eight inches (perhaps with a cavity), to single block width of about three-and-one half inches and obviously no cavity. In other words the air space (cavity) was of any width or none, tied withe to withe or not. The implication is profound. The grand idea of rigid knitted "slabs" is lost. As Wright said in contradiction to his other comments, but correctly, it was "double wall construction," that is "one wall facing inside and the other wall facing outside."[49]

Technically, Wright's first block design for Millard, the one that mimicked Griffin's, was for fitted blocks that, when properly mortared together, formed a composite block eight by eight or eight by fifteen-and-one-half inches. Second and further revisions were for tiles with edge groves. Their width was such that they lacked lateral stability and therefore were of little value in compression unless two withes were tied across a cavity, or preferably a concrete filled cavity. Lloyd reported that Ennis's tall south retaining wall of a single withe of tiles buckled a few months after installation. A single withe could carry only its dead load, no live or bending or point bearing load, and had to be laid vertically with great care. The length to width ratio for columns can generally apply to walls. These limitations were surely obvious to people who like the Wrights loved brick wall construction.

Over the years, more so than the Storer and Millard houses, the Freeman and Ennis buildings suffered from earth tremor turbulence. As well, rain, acid rain and plain dampness caused deterioration of the ill-fitting and porous concrete blocks and tiles. They became flakey, crumbled, pieces and whole tiles fell off. More troublesome, since Wright specified that, since tiles were butt-jointed, mortar not be applied at the joint and therefore water easily invaded by water and moisture. In Wright's opinion all problems were the fault to the builder and/or nature.[50] The Ennis house was seriously damaged by the 1994 Northridge earthquake; the Freeman house less so. Then in March 2005 the Ennis undercarriage started to shift after torrential rains. Earth fill behind supposed retaining walls became mud and slid under the house through the thin walls and downhill. Water continued to enter behind tiles and thousands were damaged, a thousand or so just fell off. Hidden posts (made of blocks) that held reinforced concrete beams that held floors and the motor court, as previously illustrated, shifted. The great south wall (a single withe veneer that was meant to hold tons of dirt infill), fell apart. So the house was evacuated. The Freeman and Ennis houses needed deep steel reinforced concrete caissons to shore against slippage and hopefully retard the wave effects of future quake action.[51]

A related problem. After a life of just twenty years, sixty percent of the concrete blocks designed for Florida Southern College had to be replaced. They had been handmade by students, unsupervised by Wright's office, from 1937 into the 1940s. Butt jointed dry with no mortar, the rather large blocks had half-round edge grooves for rebars. But the grout, again, did not wholly cover the rebars and rust quickly set in. And the flakey blocks absorbed moisture and crumbled.[52] (So too those for the Robert Levine house of 1948. Students made the blocks using lake water full of impurities that discolored blocks that also effloresced.) With millions of dollars required to restore the four Los Angeles houses over a period of decades (and recently just the two), and to nearly rebuild Ennis, Wright's structural textile block system was a proven failure. But only structurally! Yet, we've shown that Wright's arguments for the system had concentrated on its technical cleverness and structural superiority, less on aesthetics.

As well, it is fairly obvious that after Unity Temple Wright did not give full attention to concrete as a finish material as had Irving Gill. Wright could have—and perhaps in the 1930s may have—taken advice from Gill's experience as published in 1915.

> *Instead of using [floor] paint I mix color with the cement, usually tones of red and yellow, red and brown, or yellow and brown slightly mottled. Tempered by the gray of the cement these colors produce neutral tones that are a splendid background for rugs and furniture. When quite dry, the cement should be cleaned with a weak solution of ammonia and water, given two coats of Chinese nut oil*

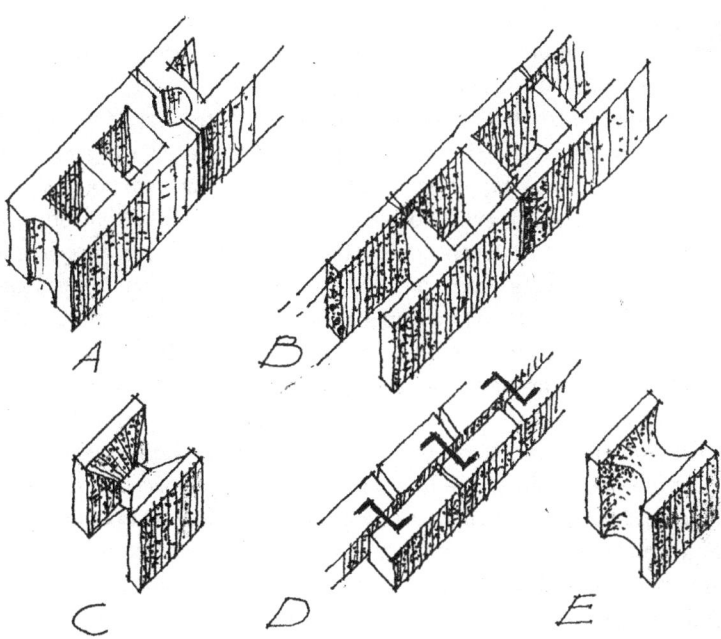

Fig. 4.8 Four concrete block shapes available commercially throughout America in 1919

Four concrete block shapes avaible in 1919 . . .

A : A standard block available in a number of size from at least 19 companys nationwide, shown is a block made by a Wizard machine, Sears, Roebuck Co.

B : T shape available from Hydra Stone Co., Chicago. 9 or 12" x 4 1/2 - 17" x 24".

C : Square or rectangular face available from S.P. Stone Co., Columbus, Ohio, or Flexo Concrete Mold Co., Cedar Rapids, Iowa. 8" x 8 or 12" x 16 or 24". Reinforcing bars could be located on four sides.

D : Brick size shown as a cavity wall with metal ties.

Available 1920. . .

E : Nel-Stone, from Nel-Stone Co., San Antonio, Texas, franchises in Los Angeles began mid-1920s. Reinforcing bars similar to C above.

With thanks to the Portland Cement Association and Sweeney(1994), 261n12-15, 209.

to bring out the color, then finished with a filler and waxed like hardwood. Well done, the treatment gives the effect of old Spanish leather.[53]

And a precedent. Prior to 1923 there were thousands of patents for hollow clay blocks or concrete blocks, some of which, as we've seen, had coved or grooved edges to receive steel rod reinforcing, A in Fig. 4.8. Moreover, four coved edges for rods and grout very similar to the Storer/Ennis textile blocks had been designed by Auguste Perret in 1920-21 for many different decorative block elements, all square, some solid, some open or (more usually) embedded with glass, for the extraordinary and well publicized church of Notre-Dame du la Raincy in Paris.[54] So, the coved or grooved edge and a glass infill were not unique to Wright but a known construction practice.

And a worry. A concrete block system called Nel-Stone (E in Fig. 4.8), was first patented in May 1920. Differences between the smooth faced Nel-Stone and the Griffin or Wright systems preclude the suggestion of appropriation by Wright or Griffin. The six inch (or larger) wall width of Nel-Stone was made of a *single* block and therefore no cavity. The recesses for steel reinforcing and cement grout (vibrated for larger blocks) were considerably larger than Wright's grooved half-wall tiles that were not vibrated. The only reason for now looking at Nel-Stone is because of a series of extant letters in the Wright Archive of the late 1920s concerning a threat to sue for patent infringement, an action not pursued.[55] In any event, by 1920 there was at least one other mass produced block very similar to Nel-Stone and readily available to consumers, C in 4.8. Nel-Stone was of little if any technical or aesthetic value to the Wrights.

The Wrights did not invent the textile block's or the tile's mechanics or chemical properties or manufacture. They initially borrowed a constructional design from Griffin and then in other ways applied—or misapplied—standard on-site techniques. Wright's patent drawing confuses the unwary. There was no application for a U.S. patent.[56] His actual contribution in the 1920s was not technical, constructional, and certainly not to commercial advancement. It was in the realm of decorative design of the exposed block faces that, with their sculptural relief, were solid, or had small openings in the pattern for air to pass through or to receive embedded pieces of glass; like Perret had done. When in place as a complete wall pattern they appeared, Wright rightly claimed, as a "delicately perforated" Persian screen.

+++

In December 1924 Walter and Marion Griffin left Sydney and traveled to the United States. On at least two occasions they visited Lloyd Wright, his Hollywood Bowl and other buildings, and Wright Senior's textile block houses in the Los Angeles area as well as Barnsdall's theatrical Hollyhock. Then in 1927 Walter wrote about those buildings for an Australian audience, his essay probably unknown in the U.S. except to close friends. In the words of architectural historian Miles Lewis, the mild and gentlemanly Griffin "did a certain amount to suggest a connection" to knitlock. Lewis quoted Griffin who had said,

> *In Southern, California [Wright] has recently carried out segmental houses on a scheme having tesseral elements of the same facial size and similar appearance [to knitlock], though quite different in structural significance, since they form cavity walls in the ordinary sense, stable because of their mass rather than through specialized columnar or concentrated supports, and there are no vertegral segments. As illustrating the expression of segmental architecture through an independent mind, his Hollywood and Pasadena projects are of much interest.*[57]

NOTES

1   Paul Kruty, *Walter Burley Griffin and the Stinson Memorial Library: Modernism Come to Main Street* (St Louis: Walter Burley Griffin Society, 2010), p. 42, drawing by Marion Griffin, p. 43.

2   On more general Australian developments in the use of concrete see Miles Lewis, *Two Hundred Years of Concrete in Australia* (North Sydney: Concrete Institute of Australia, 1988). For an insight into a slight but related episode of then current American practice (and W.B. Griffin's) see Paul Kruty, "Wood Bungalows and Burnt-Clay Cottages. Building a Model Brick House at the Panama-Pacific International Exposition," *Winterthur Portfolio* (Delaware), 40(Summer/Autumn 2005), pp. 133-152.

3   WBG to Lillian Hamilton-Moore, 15 May 1934, as quoted in Griffin (2008), p. 294.

4   Johnson (1977), p. 114,n46, more generally pp. 57-61. Walter was applicant i.e. inventor, the builder Jenkins was assignee. Application for U.S. patents for a "building component" was made on 30 April 1918 and granted 14 June 1921.

5   [W.B. Griffin], "Knitlock construction," *Australian Home Builder* (Sydney), (August, 1922), p. 73; Navaretti (1998), pp. 162-163, 194-196, for detailed view, p. 299; P.Y. Navaretti and J. Turnbull, "Burley Griffin's Pholiota," *Content* (Department of Architecture, University of Sydney, no.1, 1995), pp. 118-127. Pholiota is 21 feet square; compare with the Caretaker's Lodge for Castlecrag, Navaretti (1998), p. 258.

When as a teenager Edward F. Billson entered the Griffins' office in Melbourne as an architectural trainee in 1916 (leaving ca.1924), one of his first tasks was knitlock. He believed that the first knitlock house was Pholiota. Interview with this author February 1969.

6   Lewis (1994), p. 18; Navaretti (1998), passim. Walter died in Lucknow, United Provinces, India, in 1937.

7   [W.B. Griffin?], "Knitlock Construction," *Australian Home Builder*, 1(August 1922), p. 73, reprint without illustrations in Griffin (2008), p. 288.

8   Griffin (1927), p. 14, article with seven illustrations, reprint with one illustration in Griffin (2008), pp. 290-292; repeated in part *West Australian* (Perth), 21 January 1928, p. 4b-c; again in part Walter Burley Griffin,"Concrete Construction Conforms to Landscape in Castlecrag Subdivision, Middle Harbour (N.S.W.)," *Highways* (Sydney), 3(14 August 1930), pp. 39-40, reprint in Griffin (2008), pp. 237-239, without illustrations. Around 1927 the Griffins referred to their Segmental Reinforced Concrete Company, see <nla.pic-vn3661835>.

Knitlock roof tiles are of little significance but see Lewis (1994), pp. 21-22; Johnson (1977), pp. 58-60. Cement asbestos tiles or shingles of similar appearance to Griffin's were produced in the U.S. in the 1910s and 1920s.

9   The Natco book was produced by Rogers and Manson Company, Boston publishers of *The Brickbuilder*, a magazine devoted to clay building products and of increasing national influence, later changing its name to *Architectural Forum*, AForum herein.

Apparently Elgh (not Elge) was from Chicago where he and Roy Lippincott worked for the Griffins on the Canberra competition drawings 1911-1912; Rubbo (1998), pp. 42-43. After a couple of years he returned to the U.S.

10   Charles E. White, Jr., *Hollow tile construction* (Philadelphia: David McKay, 1919, reprint 1924), pp. 24-28, figures 55-57; on "Texture Tile" see Frederick Squires, *The Hollow Tile House* (New York: William Comstock, 1913), passim; with thanks to Prof. Miles Lewis who generously shared his research on Natco.

11   Wright (1928), p. 102. See also Onderdonk (1929), where information about the Wright buildings was supplied by Wright and Lloyd Wright or taken from Wright (1928).

12   Sweeney (1994), p. 7.

13 Of course design development preceded the preparation of construction drawings that were dated December 1922, Gebhard/von Breton (1971), p. 9.

14 As put by Gebhard/von Breton (1971), p. 34, after both authors had lengthy discussions with Lloyd; see also p. i,10,72n.20 and n.21; and Weintraub (1998), pp. 18-21. The house best visually displayed in Paul Goldberger, "Reorienting a Classic," *Architectural Digest*, (March 1987), pp. 108-115; Weintraub (1998), pp. 54-61; and Weintraub/Hess (2007).

15 As quoted in Esther McCoy, "Lloyd Wright," *Arts and Architecture* (Los Angeles), 83(October 1966), p. 24; and Sweeney (1994), p. 205; followed by a quote by Dana Hutt, Weintraub (1998), p. 54.

16 Gebhard/von Breton (1971), p. 26.

17 Gebhard/von Breton (1971), p. 26.

18 Lloyd W. to FLW, 8 February 1928, 1, copy LW Papers. Three wall sections of the Bollman house are illustrated in Weintraub (1998), p. 56, but knit-blocks are not shown. The better known of Lloyd's buildings to use concrete masonry (if not throughout) is the ornate (gaudy?) John Sowden house, Los Angeles (1926), see illustrations in Weintraub (1998), pp. 76-89; cf. Gebhard/von Breton (1971), pp. 31-35, 41; and Onderdonk (1929), pp. 112-116.

19 Lewis (1994), p. 23.

20 Figures 21, 22 and 27 in Sweeney (1994); Chusid (1992), p. 71. Former curator of the Freeman house for the University of Southern California, Jeffrey Chusid believes that the Millard house "is not actually built using textile block; it lacks the two way steel reinforcing, and the thin coffered tiles." Just so, as noted herein, but Wright did not explicitly make the distinction.

21 Figure 68 in Neutra (1927), from a photograph no doubt supplied by the Griffins, also reprinted as Figure 210 in Sweeney (1994), and elsewhere.

22 Original illustrated in Smith (1992), p. 167; Smith indicates the commission for Barnsdall's Beverly Hills house was withdrawn the following July.

23 Wright to architect Jacques Andrè in Nancy, France, 23 November 1930, Pfeiffer (1984). pp. 82-83.

24 Wright's use of the square as a planning and aesthetic determinant for two and three dimensional design, architecture and city planning is a major topic of this author's research; see the outline and geometrical analyses in Johnson (2004), pp. 3-28.

25 One patterned glass block is pictured in color in Hanks (1989), p. 29.

26 Hanks (1989), p. 9, 70. See also Pfeiffer (89-16), plates 214, 219, 224; Alfosin (1993), pp. 209-210, 234-235, 275-276; David G. De Long, ed., *FLW and the Living City* (Weil am Rhein: Vitra Museum, 1997), p. 252; Nute (1993), pp. 88-93, but it should be read in the light of other research.

27 On the evolution of surface designs 1917-1925 see Alofsin (1993), pp. 162-300; Nute (1993), pp. 88-94; Sweeney/Calvo (1984), pp. 63-78; see also Frampton (1991); and the intriguing analyses of Lionel March and Philip Steadman, *The Geometry of Environment: An introduction to spatial organization in design* (London: RIBA publications, 1971), pp. 38-79.

28 The interior of the Millard house is seldom illustrated but see Weintraub (1998), pp. 68-75; *Sunset* (Menlo Park), 197(October 1996), p. 98; and Weintraub/Hess (2007).

29 The stone is located at Llangammeret Church, Brecon; see Lewis Spence, *The Mysteries of Britain* (London: 1928 and reprints), following p. 165. The Celtic symbol was used abstractly by Wright on some of the diaper tiles on Midway Gardens. Taliesin was the name Wright gave to his houses in Spring Green and Scottsdale.

On Welsh traditions as related to FLW's maternal family and architectural philosophy see Thomas Beeby, "The Song of Taliesin," *Modulus. The University of Virginia School of Architecture Review* (1980-81), pp. 2-11; Neil Levine, "The Story of Taliesin: Wright's First Natural House," *Wright Studies*, pp. 2-27; on possible sources of FLW's knowledge of Welsh traditions see Scott Gartner, "The Shining Brow: FLW and the Welsh bardic tradition," *Wright Studies*, pp. 28-43.

On the context of Taliesin's life and poetry and some reasonable analyses of his writing see Geoffrey Ashe, et al, *The Quest for Arthur's Britain* (St Albins: Paladin, ca.1968), pp. 163, 173-175; and Nora K. Chadwick, *Celtic Britain* (London: Thames & Hudson, [1963]), pp. 101-106, and as examples only of the symbol see plates 10, 28, 432a, and Figure 22. The symbol was incorporated into the Christianized Celtic cross.

30 Frank (2008), p. 248, a much needed and long overdue research publication.

31 Van Zanten (2000), p. 74.

32 Van Zanten (2000), p. 74.

33 Dow (1899); cf. Nute (1993).

34 Nute (1993), p. 87.

35 As quoted in Nute (1993), p. 87.

36 [Report], *Architecture* (New York), (June 1900), p. 224; Nute (1993), p. 86, 185.

37 Ross (1907), p. 5.

38 Ross (1907), p. 5.

39 As quoted in Van Zanten (2000), p. 75.

40 Wright (1932), pp. 159-160, his ellipses.

41 Dow (1899), p. 17; Nute (1993), pp. 86-89, 93.

42 Siry (1996), p. 155. See also Nute (1993), inter alia; and Ernest A. Batchelder, "Design in theory and practice: a series of lessons," *Craftsman* (New York), 13(October 1907), pp. 82-89.

43 Alofsin (1993), Chapter 9 and pp. 295-306.

44 Wright to Rebori, 15 September 1927, copy Wright Archive, as quoted in Sweeney (1994), p. 118; quoted in part in Rebori (1927), p. 453.

45 Neutra (1927). The Neutra drawing was first published in Richard Neutra, "Eine Bauweise in bewehrtem Beton Neubauten von FLW," in de Fries (1926), a book that received Wright's cooperation. In the 1950s metric dimensions were changed to inches for Wright's book *The Natural House*; reprint Smith (2005), p. 8.

46 See construction drawings for the Storer house and the Griffins' knitlock nicely redrawn in Ford (1990), pp. 323-329; Edward R. Ford, "The Pioneering Age of concrete Blocks—FLW's Textile-Block Houses", *Detail* (Munich), 43(4, 2003), pp. 310-315; Sweeney (1994), here and there. See also Langmead/Johnson (2000), Chapters 8-9.

There are no drawn details of Wright's blocks in Neutra (1927). It was the first volume of Stuttgart publisher Julius Hoffman's series surveying modern architecture:

Groszstaat Architecture (v.2, 1927), and *Internationale Neue Baukunst* (v.3, 1928), the latter an edited reprint of Hoffman's *Modern Bauformen. Monatshefte für Architecture und Raumkunst* (1927). Each volume was compiled by Ludwig Hilberseimer (later to join Mies van der Rohe at Illinois Institute of Technology, Chicago), and in each book Wright's work is illustrated, including the concrete block houses.

47 Wright to L. Wright, 1 February 1928, p. 1; L. Wright to Wright, 8 February 1927, p. 1, LW Papers.

48 Wright (1927b), p. 319. When writing for architectural publications Wright tended to emphasize practical matters. In autobiographical and other essays the tendency was to write for a general though lay but impressionable audience.

49 Rebori (1927), p. 456; FLW, "In the cause . . . IX. The Terms," *ARecord*, 64(December 1928), p. 315.

50 On Wright's confessions about Millard's house see Wright (1932), pp. 248-249.

51 Alice Ormsbee Beltran, "Water in the Freeman house: a study of the block in the textile block construction assembly as the primary route for water infiltration," M. of Building Science thesis, Faculty of the School of Architecture, University of Southern California, December 2006; on damage to and reconstruction of the Ennis and Freeman houses outlined by Smith (2005), especially pp. 16-19. Replicas of textile block designs for some of the L.A. houses are available from retail outlets and on the web.

In spite of maintenance difficulties the Millard house sold for $1,185,000 in July 1996; the asking price in 2008 was $7,733,000.

52 Cf. Storrer (1993), p. 260; Kenneth Powell and Susan Dawson, "The Wright spirit; Original architect. . .", *Architects' Journal* (London), 214(November 2001), pp. 224-33; Jeffrey M. Chusid, "Preserving the Textile Block at Florida Southern College" (New York: World Monuments Fund, 2011), report dated September 2009 at <www.wmf.org>, accessed June 2012; Aaron Hoover, "Fixing Wright's Wrongs", *Coastal Contractor* (Fort Lauderdale), March 2008, pp. 1-3.

53 Irving Gill, "New Ideas About Concrete Floors," *Sunset Magazine* (Menlo Park), (December, 1915), as quoted in <www.irvinggill.com>, accessed 6 October 2008, the best online site for Gill.

54 On Perret's block designs see for instance Jean Badovici, "Notre-Dame du Raincy par A. et G. Perret," *L'Architecture Vivante* (Paris, 1923), pp. 14-15; Collins (1959); Peter Collins, "Auguste Perret", in *The Anti-Rationalists and the Rationalists*, J.M. Richard, Nikolaus Pevsner and Dennis Sharp, editors (Oxford: Architectural Press, 2000), pp. 16-25, and note the modern buildings of Denis Honegger and John Fassler; and for purposes herein Leonardo Benevolo, *History of Modern Architecture* (London: Routledge & Kegan Paul, 1966), v.1, pp. 328-330.

55 Discussed in Sweeney (1994), pp. 208-214 see various letters about it (and Lloyd's redesigned blocks), 2 September 1927 (when Lloyd first became aware of the threat) through 1932, LW Papers.

56 FLW to Lloyd, n.d. but subsequently dated "1920" perhaps by Lloyd, p. 1, LW Papers; and Secrest (1992), p. 354.

57 Griffin (1927), p. 62.

5

Wright's Fiction

Concrete: Aesthetically it has neither song nor story.
The cheapest (and ugliest) thing in the building world . . . that gutter-rat.
Concrete is a plastic material—susceptible to the impress of imagination.
— Frank Lloyd Wright, 1928, 1932

Father and son's borrowing, clearly a conscious act rather than subliminal, was and is not uncommon, certainly not in the realms of art and architecture. Wright Senior's publicly uttered justifications for the pursuit of concrete block and the value of his resulting designs, however, were not just exaggeration and bluff but entered the realm of fiction. "Fiction" is an awkward word, as most postmodern theorists reveal when attempting to transfer theoretical literary propositions and analogies to the complexities of architecture, of building buildings. It is used here in relation to what people have said, not to buildings, and in the following sense: fiction as something feigned (that is pretended, counterfeit, a sham), invented (in this instance something concocted or fabricated), or imagined (pretended or made up). Adding to what we will discover as Wright's pretensions, his more ardent observers and critics have offered other fictions about Wright's architecture in the course of accepting his words too literally, or as fact. Those fictions related to precedent, intention, history, theory, and occasionally to invention. Some were fatuous, others simply erroneous; many just puerile.

In relation to concrete blocks and tiles, an understanding of the content of those coarse verbalistic phenomena, Wright's and theirs, is followed herein by an investigation of his motivations, implied or ignored, as they illuminate intent, together with tolerable interpretative conclusions. These include Wright's stories as to why he began experimental and constructional investigations of concrete blocks in 1923. We've discovered that the major catalytic contribution was the handy information about the Griffins' knitlock invention supplied to the Wrights in 1919. That fortuitous episode was followed immediately by son Lloyd's otherwise insignificant knit-blocks. Beyond those two events answers are most intriguing.

MONOLITH

In the five decades before 1917 there was a building boom in the United States as one result of dramatic industrial and commercial developments and a concomitant increase of population in northern urban centers. As a new and fireproof construction material, reinforced concrete proved to be more efficient and flexible than conventional materials of stone and brick, of cast iron or the newly available steel, a material that so easily deforms under extremes of heat that it must be protected. Concrete was and is not cheaper than fired terra cotta but structurally superior. Except in compression as masonry, terra cotta—i.e. fired clay—has no structural value.

Around 1900 it was thought by a few people that the application of concrete would help identify in technical and aesthetic terms an architecture suitable for the new Christian century. In America it was, after all, the beginning of the Progressive Era, that great educative reaction to the previous decades that Mark Twain and Charles Dudley Warner satirically but accurately described as the Gilded Age. (It is too often incorrectly called Victorian, a period solely of the British monarchy, not of American politics or anything else.) Scientific methodologies and simple fact finding would enter town planning, social studies, housing, management (such as Taylorism), and all else relevant. They paralleled a rising pragmatic social conscious and a rapid technological development—applied science—that infused structural and mechanical engineering and construction materials. The most potent symbol worldwide of America's unfolding integration of technology with commerce, manufacturing, urbanization, and design was the skyscraper.[1]

After the 1918 Versailles treaty American Progressivism fragmented, became less potent under the impress of a reactionary conservatism. But technology continued to flourish as it satisfied the demands of industry and commerce. To satisfy the need for cheap housing for the thousands of southerners moving north and an increasing European migration to North America it was thought that advanced building construction techniques could help. We've looked at Atterbury's and Morrill's singular contributions. An interesting investigation by Schindler and Wright was undertaken in response to an invitation in 1919 by attorney Thomas Paul Hardy. That was before Schindler left Oak Park to superintend Wright's Los Angeles office. Hardy had commissioned Wright in 1905 to design a family home in Racine, Wisconsin. An academically ordered building it was completed in 1906.[2] Now he hired Wright to design eighteen fireproof "workers" cottages as a real esate scheme on another site in Racine. With Wright in Japan, Schindler, the acknowledged designer, provided plans that incorporated unadorned vertical and horizontal concrete slabs embedded with reinforcing bars bent ninety degrees so as to be continuous through the ninety degree connection of floors to walls and walls to roof. It was a common engineering practice.

As with the Tokyo Hotel, if a building was an inflexibly bound concrete unit with all parts tied together by steel reinforcing rods, then in an earthquake it would act as a solid block riding on moving earth. It is not known if this was the motive for houses that were to be in a place where earthquakes were of little consequence. More probably it was for fireproofing. That aside, the construction details were in

Fig. 5.1 Monolith Homes commissioned by Thomas P. Hardy as a prototype for a site in Racine, Wisconsin, 1919

fact not unique but common. Since something like a system was involved and the houses appeared special, Wright probably did not submit an application for there is no record of a grant by the US. Patent office. Other aspects of Monolithic Homes, as the architect named them, have been described confusingly as follows.

> The continuity of surface resulting from the pouring process was countered [?] by Wright's squaring of the wall surfaces into a grid. Each unit of the grid contained a screen-covered square, a "concrete separator" [?], at a corner, which was intended to provide ventilation to the slab and a pattern to the [otherwise] monotonous concrete surface.[3]

At the corners between vertical wall slabs there was to be floor to ceiling windows, the glass held by a ladder of horizontal wood members. Fig. 5.1 above is a perspective drawing of "The Worker's House"; Schindler signed his name in the drawing title block "RMSchindler / First Sketch / Oak Park, Illin / July 6th 1919". Other drawings were dated September 1919.[4] No doubt the design was based on discussions with Wright between trips to Tokyo, or via sea mails. Also, Schindler's pencil drawings were executed before Wright's return to Spring Green in October

1919. Then in a sketching frenzy he added bunches of ornamental embellishments to an elevation drawing still held by the Wright Archive.

Monolith Homes remained unbuilt but the design did herald aspects of the California concrete block houses to follow: plain rectilinear walls, corner windows, square proportions, a boxey appearance, and flat roofs, all easily identified with Schindler's maturing design ability.[5] Further, Monolith's basic parti in floor plan and in section was transferred in 1920 to a set of "Terrace Houses" for Aline Barnsdall. They were to be on the edge of Olive Hill beside the northern street as a line of small shops and studios for her art colonists. However concrete was discarded as the construction material. Although nothing else changed in the design, walls of brick were to be coated with stucco and plaster, floor and roof joists of wood. After later modifications they were called Residences C, D, E and F. A complete set of construction drawings were prepared by Schindler, but the Terrace Houses also were not built.[6]

WORDS

Wright's first published verbal descriptions of the textile block construction system were in two publications only weeks apart in 1923. One was within a short twelve page pamphlet about problems he saw in the centralizing effect of modern urbanization, the bad example Los Angeles. He say an opportunity to participate in the creation of a new urban environment for post-war, post-earthquake Tokyo. The pamphlet was about the benefits of a future with decentralized towns. Perhaps Wright was intrigued, momentarily, by G. Gordon Whitnal's call:

> *not another New York, but a New Los Angeles. Not a great homogeneous mass with a pyramiding of populations and squalor in a single center, but a federation of communities co-ordinated into a metropolis of sunlight and air.*[7]

In September 1923 Wright outlined for Louis Sullivan the concerns expressed in the essay:

> *I am opposed to the tall buildings [Wright said,] ... I have written something outlining these views ... and I am sending it to Tokyo to try and head off the propaganda which will try to rebuild Tokyo as a modern American City.*[8]

A shortened and modified essay was sent off to Tokyo where it was published in the *Japan Advertiser* under the title "Why Perpetuate the Skyscraper".[9] Yet during 1923 and into 1925 he proposed three, perhaps four, high rise office buildings in the hope of enticing clients. One was for a thirty floor reinforced concrete building with cantilevers balanced on "pylons" that were structurally more extensive and daring than the hotel. The interior and exterior walls were to be copper and glass. Apparently around May 1923 it was first offered to some businessmen in Los Angeles and ignored.[10]

Anyway, the pamphlet entitled *Experimenting with Human Lives* followed upon the well publicised fact that the Imperial Hotel, built mostly on firm ground, less on loose and soggy landfill called "cheese", had survived a major earthquake in April 1922; and so had Wright. ("I myself, [he said] caught in the upper story [of the Imperial Hotel]. . . . I was dripping with perspiration myself, knees none too steady".[11]) Seventeen months later on 1 September 1923 the building survived the great Kanto earthquake and subsequent fire.[12] There were over ninety thousand deaths and the Tokyo of wood buildings was destroyed, seventy-five percent gray ash. The hotel remained standing with relatively minor damage, a credit to Chicago engineer and construction company owner Paul F.P. Mueller and architect Dankmar Adler. They had designed foundations for the Adler and Sullivan Auditorium Building in Chicago of 1886 to 1889. The foundations were special because the building was to sit on soft clay of a depth of 100 feet, a soil spungy when wet like the soft infill beside Tokyo Bay. Wright was familiar with details of the Auditorium because he was hired as a draftsman for it. In and round 1918 Wright hired Mueller who joined him in Tokyo. Yet, Wright has said he consulted a dozen of so engineers including the local office of Truscon Steel.[13] Wright made much of the hotel's survival to audiences in Los Angeles, Chicago and Tokyo, and to wider North America, all in publications. He even asked Louis Sullivan to write a piece.[14] Little was offered in that publicity about the engineering contribution.

However, engineers apparently were not involved with the Arinobu Fukuhara house in Hakone for which Wright was the principle designer. It was destroyed by the Kanto earthquake, a fact reflected upon by Masami Tanigawa: the story of the Imperial Hotel's survival "has been repeatedly told, but nothing about this Fukuhara House".[15] But there were other troubles for the hotel.

The Imperial Hotel suffered other serious problems, technical rather than aesthetic. Much of the building's interior and exterior ornamentation, its "finished surfaces", as Wright put it, were made of a local *oya-ishi* or oya stone, actually a tuff and easily cut when green. There were two problems with the tuff. First, it was composed of bits of a multicolored material that was coarse and pitted with deep pocks. Wright's patterns in the stone, most finely cut in low relief, were/are very difficult to read, detail faint, indistinct. If this registered with Wright during selection of materials he still persisted. Second, when air-cured the exterior tuff nonetheless disintegrated in the city's increasingly acidic atmosphere; so too some of the exterior fired clay ornament.[16]

By often crude patchwork the hotel managed to survive more earth quakes and tremors, cyclones, increased acid rain and World War 2. Perhaps too hastily, in the opinion of many people, it was demolished in 1968. The two storey entry and lobby as a discrete building section was moved to Meji Mura Village near Nagoya where reconstruction was complete in 1976. There it became a great participatory sculpture.

The Imperial Hotel was the major design commission undertaken during the seven years before detailing his California concrete blocks. It was also a building of monumental size and of reinforced concrete construction. In his outline of engineering specifications he insisted on normal practice: the hotel walls be "cast solid . . . within thin shells of facing material"., concrete floor "slabs" be "continuous side to side of building with continuous reinforcement", much like Midway Gardens and Monolithic Homes.[17] Other conditions ensured the structure would be a series of monolithically constructed discrete units of reinforced concrete that would independently ride on the waves of earth when attacked by a quake. With the success of the hotel's engineering, why were the same structural principles, the same care, not applied to the four Los Angeles houses?; especially to the elongated Ennis house sitting awkwardly on soft earth while during rain storms trying to resist the pressure of water-saturated backfill soil. After all, it was designed a few months after the Tokyo hotel survived the Kanto quake.

Wright was not yet done with Japan. In the wake of the publicity attached to the hotel's proven endurance he attempted to insert ideas into the ongoing debates as to how to reconstruct Tokyo. He was tempted to do this perhaps as a reaction to adulations by his Tokyo architectural colleague Arata Endo. After witnessing the Kanto quake, in 1923 Endo wrote to his "Lieber Meister" not only in praise but revealing a knowledge of Wright's dismal professional reception in Southern California. Endo invited him to return east and again set up a professional office in Tokyo . . .

> *Now your chance is here. You will be received here now with admiration and appreciation—late, yes, but not too late. The whole city is at your disposal. Your work here has been prepared for you. You will have more appreciation now than in America. Therefore you had best come here where it is more worthwhile to plant your footsteps than in Los Angeles—don't you think?*[18]

In his 1923 submission to people in Tokyo and in the related pamphlet, Wright envisaged a few buildings of about seven stories, not centralized but scattered about the revived city. As noted above, there would be no skyscrapers as found in America. For housing of three storeys or less he suggested the following miraculous technology:

> *Concrete blocks with steel interlaced two way and joints poured [i.e. packed with concrete grout] are advocated because a really flexible wall is thus secured, [it is] less liable to damage from flexure or torsion [twisting] and one that would show no cracks.*[19]

Following his experience with the structural efficacy of the hotel he added that "the foundations should be flexible; that is, concrete piles poured into holes made in the . . . ground and a reinforced linear slab built on the mud as on a cushion".[20]

No more was said on the subject. Nor were textile blocks presented in the *Wendingen* publication of 1925 edited by H.Th. Wijdeveld and in which Wright was so fully involved. Then in a typescript prepared in 1927 for his old friend Andy Rebori, a Chicago architect who was then an editor of *Architectural Record* in New York. Wright provided slightly more detail that Rebori incorporated into what became a glorifying article.

> *Tokio[`s] . . . principal residence and business areas [Wright said to Rebori, should be] restricted to two or three stories.*

> *The first story and if three . . . the second should be hollow or double walls made of reinforced concrete blocks—interlaced by 1/4" steel rods every 16", vertically and horizontality. These walls should extend to the window sills of the top story or be used as a railing for balconies or for the top story windows. . . .*
>
> *Within the outer walls and resting on the upper floor-slab should be set and securely doweled the lightest possible scientific timber framing of walls and roof of top story . . . .*[21]

The typescript prepared for Rebori contained some of the text prepared for the *Experimenting* pamphlet.[22] Wright's next public comment about textile blocks was in the following year.

While the Ennis house was nearing completion in 1925, Wright affirmed its modest technology as intimately of the region, the blocks "made from the gravel of decayed granite of the hills easily obtained there and mixed with cement and cast in molds. . . ".[23] Wright also said tiles for the houses were "made of various combinations of the decayed granite and sand and gravel of the sites". No doubt it was the soil's organic impurities that caused most blocks to begin disintegrating after only a few years. Substitute cement, sand and water for mud, grass and water and the formula is similar to sun dried adobe, the ancient and universal building material. In some cultures dung and/or animal hair are added. However, in Wright's mixture there were no brightly colored aggregates to expose as Atterbury and Arts and Crafts people had so delightfully fashioned.[24]

When writing a series of essays on architectural theory (of sorts) for *Architectural Record* in 1927 and 1928, and as later amplified or restyled while preparing an autobiography during the years 1928 into 1931, and while the Arizona Biltmore was under construction in Phoenix, and while another hotel to use textile blocks (the ill-fated San Marcos near Chandler, Arizona, 1927-1929) was in preliminary planning, Wright wrote in difficult prose, like bursts of thought, about reinforced concrete and the imaginative process leading to the creation of the textile block system. He offered a mix of a little theory, a lot of sentimental justification, much obfuscation, a deal of sophomorically effusive prose, and half-truths as the following examples attest.

> *I am writing this on the Phoenix plain of Arizona. The ruddy granite mountain-heaps, grown "old", are decomposing and sliding down layer upon layer to further compose the solid of the plain. Granite in various stages of decay, sand, silt and gravel make the floor of the world here.*
> *Buildings could grow right up out of the "ground" were this "soil", before it is too far "rotted", cemented in proper proportions and beaten into flasks or boxes—a few steel strands dropped in for reinforcement.*
> *Cement may be, here as elsewhere, the secret stamina of the physical body of our new world.*
>
> *And steel has given to cement (this invaluable ancient medium) new life, new purposes and possibilities, for when the coefficient of expansion and contraction was found to be the same in concrete and steel, a new world was opened to the Architect.*[25]

Written for a professional journal in 1928, the above adequately displays his manner of presentation to fellow architects. Separating out some of the more coherent prose he found that concrete ...

> A valuable partnership in materials in any case more congenial ... than steel alone for he [the architect] can do more richly with flesh and sinews than he can with sinews and bones, perhaps. ...

> Here we have reinforced concrete, a new dispensation. A new medium for the new world of thought and feeling that seems ideal: ... freedom from the imprisonment in the abstract in which tradition binds us. Democracy means liberation from those abstractions, and therefore life, more abundantly in the concrete. This is not intended as a pun. It happens to be so literally, for concrete combined with steel strands will probably become the physical body of the modern civilized world.[26]

Among other observations and comments he also wrote that:

> The Machine is an obedient, tireless fabricator of a non-sentient product. A shaper and drawer of steel, a weaver of fabrics—"casting" forms continually in every material solvent by fire or water. ...

> Imagination is so intimately related to sentient perception—we can not separate the two. Nor need do so.

> [Of things mechanistic] let us take one that is both a chemical-process and casting—concrete. ...

> [Concrete] is still a mass material taking form from moulds, erroneously called "forms".

> Unity Temple at Oak Park [1905-1908] was entirely cast in wooden boxes, ornamentation and all. The ornament was formed in the mass by taking blocks of wood ... combining them with strips of wood, and, where wanted tacking them in position to the inside faces of the boxes. ...

When Wright came to presenting ideas about concrete block construction the words were much too general yet infused with a sense not only of professional propriety but theoretical joy. But typical of so much of his published writing Wright tended to talk down to his non-professional *and* professional audience, patronage barely hidden. For example: "Mechanical moulds" are taken to a building site, he said, where

> gravel and sand abound. Cement is all else needed, except a few tons of 1/4 in. commercial steel barbs, to complete a beautiful building. ...

> The ground is soon covered with slab-blocks, [then] it is all a matter of reading the architect's diagrams, which is what his plans now become. ... [S]howing where specific blocks go is like counting stitches in the "woof" and threads in the "warp". Building is a matter of taking slab-block stitches on a steel warp.

> *So, livable building may be made of monomaterial in one operation! . . .*
>
> *Plastering? None. Carpenter work? None. Mason? None. "Form" work" None. Painting? None. Decorations? All integral, cast into the structure as designed with all the play of imagery known to Persian or Moor.*

A half truth is half untrue. Shortly afer marriage in 1928 Wright began writing an autobiography designed almost wholly for a popular audience of future clients and for a place in history. It included descriptions of the concrete blocks, now using the word "textile".

> *Gradually I unfolded the scheme of the textile block-slab house gradually forming in my mind since I got home from Japan [in 1922]. . . .*
>
> *The concrete block? The cheapest (and ugliest) thing in the building world. It lived mostly in the architectural gutter as an imitation of "rock face" stone.*
>
> *Why not see what could be done with that gutter-rat? Steel wedded to it[,] cast inside the joints[,] and the block itself brought into some broad, practical scheme of general treatment[,] then why would it not be fit for a phase of modern architecture?*[27]

He glanced at the rising influence of the mechanistic European modernism then appearing often in American architectural magazines. They exerted an increasing influence nation wide, especially on students in schools of architecture. But in the autobiography he continued to speak more to the amateur or lay audience than to wiser professionals, engineers included.

> *There should be many phases of architecture as modern [Wright said]. . . .*
>
> *Concrete is a plastic material—susceptible to the impress of imagination. I saw a kind of weaving coming out of it. Why not weave a kind of building? Then I saw the "shell". Shells with steel inlaid in them. Or steel for warp and masonry unit for "woof" [weft] in the weaving. . . . [A]ll such units or block[s] for either weaving or shells to be set steel-wound and steel-bound. Floors, ceilings, walls all the same— all to be hollow. . . .*
>
> *Lightness and strength. Steel the spider, now spinning a web within the plastic material. . . .*[28]

(Surely the spider would have been the construction worker, steel the web.) We've learned that what he, the "Weaver" (he called himself), here described rather floridly if not meretriciously, and at times quite inaccurately (including metaphors) was not new. His words simply transposed clay (terra cotta) brick or clay hollow block building practices of a cavity—or non-cavity—wall to concrete blocks and then to tiles.

In phrasing and language, however, his autobiographical and theoretical musing from 1927 to 1931 about the blocks impute a heady implication that they were his invention, born out of a desire to legitimize concrete through his talent by suggesting the creation of a new method of construction. This is all too clear in the quotations above. Indeed, in 1927 he said that his and Lloyd's "concentration" had been "on invention".[29] Here he confirms Lloyd's collaboration. But that statement was otherwise erroneous, another fiction. His description, still obviously for a popular lay audience, continued.

> *All we [he and Lloyd] would have to do would be to educate the concrete block, refine it and knit it together with steel in the joints and so construct the joints that they could be poured full of concrete after they were set up and a steel-strand laid in them. The walls would thus become thin but solid [now not flexible] reinforced slabs and yield to any desired form imaginable. And common labor could do it all.*

To Wright "common labor" meant non-unionized labor because he expressed a hatred for unions. (Lloyd gathered Mexican migrants off the streets.) That last paragraph described a concrete wall where the normal outer wood formwork either side is replaced by the thin textile concrete tiles. There is no cavity. He had used a similar reinforced *in situ* wall for the Tokyo hotel where exterior formwork was not wood but masonry. The outside wall was "laid of [clay] bricks", Wright told us, "that formed a 2 inch wall with 4 inch spurs with in the wall 4'-0" on centers. The inside face ... was laid up with 4"x12" [hollow clay] tile blocks with dovetailed grooves...".[30] The cavity between the two masonry walls of single withe contained steel reinforcing rods and then filled with poured concrete.[31] In Fig. 5.2 "shell" refers to the inner and outer withe of clay masonry, not the complete wall. He called it "double shell construction".[32] After discussions with Wright, Louis Sullivan described it so:

> *wooden forms were dispensed with; the outer layer of specially notched bricks, and the inside layer of hollow bricks, serving as such. In the cavity between, rods, vertical and horizontal, were placed, and the concrete filler, the wall thus becoming a solid mass of varied materials, into which the floor slabs are so solidly tied....*[33]

The "specially notched bricks" were probably those with dovetailed grooves that Griffin had designed for Natco.

Concrete as a wall filler or a stabilizer for a masonry arch was, as previously noted, well known since its extensive application in Roman antiquity and after. Fig. 5.2 is a fair illustration of typical Roman wall construction. When the cavity is "poured full of concrete" it obviously cannot allow "continuous hollow spaces" in walls as Wright now and then had described. "Floors, ceilings, walls ... all to be hollow", he said, and ...

> *We [he and Lloyd] would make the walls double[,] of course, one wall facing inside and the other wall facing outside, thus getting continuous hollow spaces between, so the house would be cool in summer, warm in winter and dry always.*[34]

This was a reflection of what had been put into practice by the Wrights.

Fig. 5.2 Pictorial explanation of "double shell" wall construction as applied to the Imperial Hotel in Tokyo, 1917-1923

From the 1890s through the 1930s Wright argued that machines must be servants to the artist-architect, they liberate him (and the builders) from the medieval caste of craft laborer. Leaving spiders and webs, his example was the analogy of the modular warp and weft strands in woven carpets wherein "lies the primitive [i.e. an original] principle of standardization: serving the imagination full well", he said. Further, the machine . . .

> *serves the spirit [and] its mechanics disappear in the glowing fabric of the mind—the poetic feeling of the artist weave. . . .*
> *An artist is sentient [Wright said]. He is never fooled by brains or science or economics. He knows by feeling—say instinct—right or living, from wrong or dead.*
>
> *He may not, however, have the technique to make his "knowing" effective and so remain inarticulate. But it is his duty to know, for his technique is what makes him serviceable to his fellows as artist.*

In other words, when the architect understands "the technique of the Machine as the tool" he will become articulate.[35] When composed by three-dimensional, orthogonal planning and a structural module (3-D warps and wefts?), all shall be integrated to the whole by the architect's will, by "the impress of human imagination", he said;[36] the designer's gift, we say.

Wright's theoretical justification was suggestively if imprecisely drawn from well known technological conditions that he attached in a rather strained, almost obscurant visionary manner to aesthetical, theoretical, and philosophical motives.

His practical description continued. "I had used the block in some such textured way in the Midway Garden upper walls".[37] In 1915 Wright said of the just installed blocks, Fig. 5.3, that at Midway Gardens "Concrete has been cast in wooden moulds with [a] new technique and a permanent character given the most ornamental as well as to the most fundamental forms".[38] (Wood molds were also used in 1915 for the German Warehouse attic tiles.) We've seen that it was not a new technique. Anyway, to the back of the fairly thick cast concrete elements ($2^3/_4$ inches or 50.8mm) were applied as tiles, cemented to the surface of brick walls that together formed a double masonry wall without a cavity. Visually the tiles presented a geometric pattern in low relief ("textured") that was employed much as a medieval tile diaper. Diapers were common, many architects employed them including Sullivan during Wright's tenure. In turn Wright initially used them in a pattern of colored terra cotta tiles on the Winslow house at River Forest, Illinois, most probably completed in 1894.

While Midway Gardens, the German Warehouse and the Imperial Hotel were the immediate technological and practical experience, there was an earlier design trial. That was the first design for an inexpensive Women's Building and Neighborhood Club for Spring Green's annual Inter-County Fair of 1914 (see Fig. 5.4). Wright designed a building that in plan had three articulated areas. One was a larger a two-storey high display room with a shallow balcony on four sides. That room was connected to single storey entry, stairs and toilets that were next to a children's play yard and related resting room. The structural and building materials are unknown but the probable budget would suggest wood studs and joists. Some kind of large patterned tiles were proposed for the exterior surface of the largest volume. We don't know if they were to be colored, textured, or in relief. The perspective drawing hints that in design and size they were very similar to diaper tiles for the Coonley house at Riverside, Illinois (1906-1908), rather than those at Midway Gardens designed just months earlier. Little else is known but a ledger at the bottom of the tiles suggests they were to be applied to the surface, as they had been at Midway.[39]

Wright continued ruminations about the Midway tiles. If "I could eliminate the mortar joint", he said, "I could make the whole fabric mechanical". How butted, mortarless joints would mechanize a labor intensive system is a mystery; so too the suggestion that such a manoeuvre would "do away with skilled labor", his bugaboo. When Wright evaluated the construction process for the Millard house he claimed there was "no organization", that the molds were prepared "experimentally". He and Lloyd had picked up "Moyana" men off Los Angeles streets, and started making and setting blocks, "The work consequently was rough done and wasteful...". Elsewhere he referred to "Monyana men" and "Monyana land": Mañana soon enough for

Fig. 5.3 Midway Gardens in Chicago, Illinois, 1913-1914, a view of the north belvedere in 1914

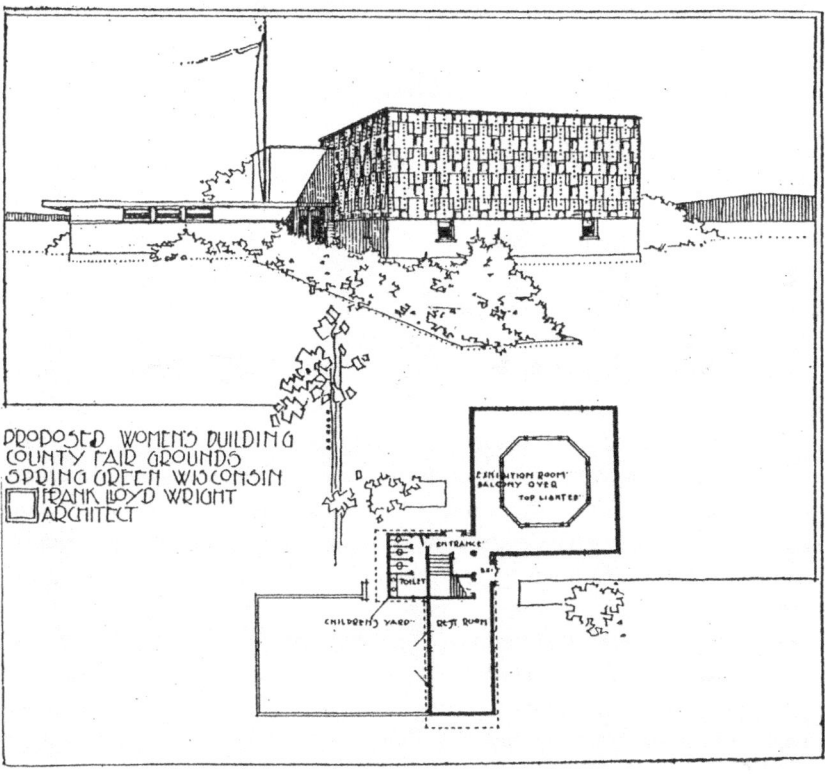

Fig. 5.4 "Proposed Women's Building, county fair grounds" near Spring Green, Wisconsin, a preliminary design of 1914

Mexican laborers? He went on to say, "All the difficulties met with were due to poor workmanship",[40] difficulties not with Wright or his detailing of the blocks but a result of Lloyd's lazy supervision of construction. For the Freeman house in 1923 Lloyd initially employed skilled builders, but after a couple of weeks they quit. So he resorted to mañana men.

Regardless and undeterred, in 1924 while houses were under construction Wright ventured to say that the textile blocks were "the beginning of a constructive [sic] effort to produce a type that would fully utilize standardization and the repetition of easily manhandled units".[41] It was a common expectation, one fulfilled by others over past centuries. The better example is ordinary concrete blocks that could have been purchased from a local shop and then a patterned face glued on. Anyway, he went on to say that "Standardization *was* [he emphasized,] the soul of the machine, and here [i.e. the four houses] the architect was taking it as a principle and 'knitting' [his quote] with it...."[42] (Knitting as in knitlock?) The analogies and language fail to impress the knowledgeable layman, the builder, engineer or architect.

Further, Wright suggested that in large measure he drew theoretical sustenance from the plain indigenous architecture of the Southwest, that it was a fertile resource, one to be modernized and enriched. We will return to this subject. For now we can record that in 1924 he said again that a building "grows as a part of its site", it should be "not planted on it". He muddled on: "[T]his sense of the indigenous thing, ... is a first condition of any successful treatment of any problem in the modern sense—as it was the secret of success in all great old work ... subconscious then".[43] That is to say, intuitive then. In a later essay he said:

> *The straight line, the flat plane, now textured [he emphasized], the sense of interior space coming through the openings all to be woven as integral features into the [concrete tile] shell. The rich encrustation of the shells visible as mass, the true mass of the architecture....*
>
> *Genuine "mass" in this sense will always be modern. A pity we ... neglect indigenous riches at any point.*

He added that he "drew my son Lloyd into this effort".[44] As we've seen, Griffin's knitlock and Lloyd's work had attracted father and then father and son collaborated.

Wright was of course primarily concerned with the visual qualities of the textile blocks, their "rich encrustation" as it enlivened rectilinear mass. Yet, since the technology he settled upon proven flawed, it would have been easier to follow something similar to what he had done at Midway Gardens. That was to cement (glue) ornamented tiles on the surface of ordinary concrete blocks or terra cotta bricks. It would have achieved the same *visual effect*. We can get a sense of this possibility when viewing illustrations herein. But that was not his purpose, his intention.

Suffice to say, nothing could have been less mechanized or less technically efficient than the making and installation of blocks, of double-wall tiles, especially as the process was not under shop supervision but handled by unskilled labor in the field: it was an ancient process.

It is doubtful that Wright sincerely believed his textile block/tile system could compete with industry; especially the pre-cut mail order kit houses of companies like Gordon Van Tine or Montgomery Ward or Bennett or Sears, Roebuck. The pioneer kit company Aladdin, begun in 1906 in Bay City, Michigan, sold houses, garages and summer cottages and then joined with suburban developers and company town planners. A single house could be built in about two weeks. Aladdin alone accounted for 2.4 percent of all housing starts nation-wide in 1918. Also, Wright must have known that in 1918-1919 Sears offered two Wright School kit houses designed by his follower, Chicago architect John Van Bergen. Nonetheless Wright's words in promotion of his system always conveyed the impression that it was not only cheaper but more quickly built.[45]

PROBLEMS

Effectively, textile blocks were hand-made prefabrications of dry slurries (one part cement, four parts sand and grit and very little water) placed in molds, wood for Unity Temple, Midway Gardens, German Warehouse and Millard, later pressed metal. When firm but not exactly set, the blocks were hand removed from the mold and then lugged up ladders to a mason. Wright admitted that construction was "clumsily home-made".[46] This is verified in photographic views of the houses and Phoenix hotel when under construction. In fact quite often many courses of exterior wall tiles were laid up before their interior mates (or vice versa), making installation of the steel ties across the cavity (if any) difficult: or there were no mates! And contrary to his enthusiasm, the rectilinear geometry of the blocks and their manner of connection and reinforcement restricted a building's three dimensional shape to the orthagonal. Those problems mock mechanical analogies.

It is useful to compare Wright's technology with Atterbury's rather than Morrill's. If we assume an Atterbury prefabricated wall panel is, say, four by eight feet, it has four edges. If we assume Wright's textile block to be one foot square, then 64 (32 each side of a wall) are required to match one of Atterbury's panels. Each of Wright's blocks needs be cast and has four edges. With 64 blocks there are 256 construction joints, therefore thousands of joints for each building, each joint a serious technical, constructional and installation problem. (Nearly all construction drawings for any building are about resolving the application of materials and their construction joints and the joints between different materials.) Each block and joint is susceptible to a multitude of vagaries during manufacture and construction, and afterward to nature's whims in the jarring forces of wind, rain, sun, gravity, earthquakes, and then to the acts of a building's inhabitants.

At a construction site each textile block had to be set beside its other, to be threaded with steel rods two ways and across a cavity, grouted, and in other ways manhandled. Like laying a brick wall, the process was indeed a continuation of medieval labor using ancient construction methods. Atterbury's hollow panels were factory built and set up on site with four joins each.

It can be said with some certainty, therefore, that Wright was not writing to be correct or even accurate, but to impress and persuade readers who ever they might be. Therefore he was not concerned with technological efficiencies or the practicalities of a machine aesthetic. He did, however, convey to audiences, lay and professional, a theoretically avant garde position and suggested a technology and appearance of modernity, a new aesthetic. In practice, however, he ignored facts and commonplace; his words almost always fictional postmortem accounts or justifications. Yet, knowing full well their serious limitations, obvious when again used for the San Marcos-in-the-Desert project and by McArthur for the Arizona Biltmore Hotel,[47] Wright nonetheless pursued to construction a variety of concrete block systems until death in 1959. Perhaps the motivation for his pursuit resided in a reflective and unfulfilled lament of 1928 . . . perhaps: "As a plastic material [he said] . . . there lives in concrete a great aesthetic property, as yet inadequately expressed".[48] Surely he meant it was not adequately expressed. In any event, was he referring to his own just completed work?

Did the laboriously inefficient construction method justify the appearance or the experience? Philosophically, pragmatically it didn't—and does not—matter. A Japanese draftsmen Wright had invited to Los Angeles in 1923, Komeki Tsuchiura, was succinct and correct: "He [Wright] was an artist. He wanted to do what he wanted to do". Terry Patterson put it this way: Wright "focused on the properties he thought potentially beautiful, to have violated others in lieu of larger goals". As to the surface designs Anthony Alofsin believed "the lowly, cast concrete block had achieved an unprecedented expressive force".[49] To Neil Levine's art-historical thinking Wright was "reinterpreting the traditional ornamental textures and patterns of earlier regional architecture".[50] Only Tsuchiura remained intellectually aloof of theoretical and practical vagaries.

Over a period of seven years Wright employed concrete blocks in designs for hotels, houses, medical clinics, fraternity houses, and so forth, all about the U.S. But those later efforts 1924 to 1929 followed the ripe moment when conception and application seemed most appropriate for a modern adobe in the Southwest in postwar America and post-revolutionary Mexico.

Fictions in Wright's explanations are intentional stories to divert our attention from material reality to theoretical stuff. His words conjure hazy romantic, visionary images that carry fascination in parallel with reason and trust. Pragmatically, the ultimate test is the clarity and value of the result. Those of us who have experienced the four houses believe them to be, well, beautiful, satiating, even mystical, most notably their interiors. The camera lens is equally satisfied as hundreds of pamphlets and books and a few films and DVDs attest. In the cause of enjoyment, pedantic analysis is, well, pedantic. In fact Wright proved himself wrong: concrete joined to talent can sing a song. And finally he was correct: the textile block's flawed "mechanics disappear in the glowing fabric of the mind". The four houses are his "homage to Imagination", they "serve the spirit".

NOTES

1 Epigraph, Wright (1928), p. 99, Wright (1932), p. 241.

The vital impact of Progressivism on FLW is one aspect of this author's continuing research.

2 Little is known of Hardy. A University of Michigan law graduate, class of '01, he apparently practiced real estate law, *Racine Journal News*, 7 March 1914, p. 15.

3 Alofsin (1993), p. 372,n.28. The Monolith house design, not the technicalities of the concrete process, was patented by Wright on 18 August 1919.

4 Riley (1994), p. 201.

5 Wright entitled the project a "Workmen's Colony of Concrete Monolith Homes" for Racine, Wisconsin; Riley (1994), p. 201, site plan p. 329; Gebhard (1971), p. 38; and on the Terrace Houses and Monolith see Smith (1992), pp. 111, 115, that includes illustrations of drawings by Schindler, Lloyd and Wright Senior.

6 Smith (1992), pp. 112-115, see construction section drawings.

7 As quoted in Mark Girouard, *Cities and people. A Social and Architectural History* (New Haven/London: Yale University Press, 1985), p. 370.

8 FLW to Sullivan, 26 September 1923, copy Wright Archive.

9 FLW, "Why Perpetuate the Skyscraper", *Japan Advertiser*, 9 January 1924, p. 7, the first issue after the earthquake, with thanks to Siry (2008), pp. 99, 105.

10 FLW, "In the cause of architecture. VIII sheet metal and a modern instance", *ARecord*, 64(October 1928), pp. 336-341.

11 Wright (1923b), p. 45.

12 Best contemporary account of the Imperial Hotel is Wright (1923a), pp. 39-46, plates 1-14; and Wright (1923c). On his reasoning as to why it survived the April 1922 and September 1923 earthquakes see Wright (1927c), pp. 61-66.

13 Siry (2008), pp. 96-98. Truscon was sold to Republic Steel in 1937.

14 Sullivan (1924), pp. 113-117; cf. Siry (2008).

15 Tanigawa (1908), p. 8.

16 Brownell/Wright (1938), pp.129-130, on the hotel p. 124-139; Alofsin (1993), pp. 295-297, 299-300, 368,n40, 373,n37-38. The loss of detail and obvious deterioration on the exterior by the 1960s is best illustrated in James (1968), throughout; and Pfeiffer (14-24), pp. 30-35.

17 Siry (2008), pp. 85-87.

18 Endo to Wright, 8 September 1923, as quoted in Pfeiffer (1984), p. 36.

19 Wright (1923b), p. 8. Reprinted and expanded with the same title, Chicago: Ralph Fletcher Seymour, [1923]. A shorter version was published in *FLW On Architecture. Selected writings (1894-1940)*, Frederick Gutheim, editor, (New York: Duel Sloan & Pearce, 1941, reprint New York: Grosset & Dunlap, 1941), pp. 97-99, where the blocks are excluded.

20 Wright (1923b), p. 8.

21 Typescript Wright Archives, as quoted in Sweeney (1994), p. 54. This text is similar to an expanded version of the initial *Experimenting* publication, probably edited after 1923, as found in Pfeiffer (1992), pp. 169-184, blocks text p. 171 and 172.

22 An outline of the arguments against the steel skyscraper was also Part VIII, pp. 19-20.

23 Wright (1925), p. 48; as quoted in Rebori (1927), p. 454.

24 Pamphlets on exposed aggregates were plentiful, one in colour and readily available was Universal Portland Cement Co., *Concrete Surfaces* (Chicago: the company, 1913). On the removal of the exposed aggregates on Unity Temple's exterior walls see J.H. Hunderman, S.J. Kelley, D. Slaton, "Unity temple: investigation and repair" in *Structural Repair and Maintenance of Historical Buildings II*, (1991), pp. 227-235.

25 Wright (1928), p. 99.

26 FLW, "In the cause of architecture". Part III. Steel", *ARecord*, 62(August 1927), p. 165.

27 Wright (1932), p. 241, part of a topic that, restyled, reiterates comments made in his 1928 series. The essays were serialized intermittently in *ARecord* under the series title "In the cause of architecture" that began May 1927 and concluded December 1928, each year a separately numbered series.

28 Wright (1932), p. 235, much of which paraphrases Wright (1928). "Woof" is English, warp is American English for the threads held by a loom; a shuttle carries weft threads normal to the warp.

29 As quoted in Rebori (1927), p. 455.

30 Wright (1923b), p. 45; Wright (1923c), p. 130.

31 The office drawing of the detail is found in color in Smith (2005), p. 8.

32 Wright (1927c), p. 64, illustrated. The Wright office detail drawing is illustrated Siry (2007), p. 88.

33 Sullivan (1924), pp. 113-117.

34 Wright (1932), pp. 241-245, a paraphrase of FLW, "Building Adventure in Modernism", [American] *Country Life*, 56(May 1929), pp. 40-41.

35 Wright (1932), p. 242.

36 As quoted in Onderdonk (1928), p. 8.

37 The example is from Wright (1927a), pp. 478-479. This can be usefully compared with Wright (1925), pp. 49, 52, 57; and Wright (1927b), pp. 319-321.

38 FLW, "In Response to a Request From the Editor for an Article on Midway Gardens", *National Architect* (New York), 5(March 1915), p. 118, a rather belligerent reply; Alofsin (1993), p. 150, 359,n68; Kruty (1998), pp. 128-131, molds p. 159,n44; photographs of precast concrete finials, figurative sculpture, and wall tiles are best illustrated in Kruty (1998), preamble and passim, and some wall tiles with their backing of a withe of bricks can be seen in a ca.1949 photograph (p. 51), perhaps the building's last remnant after being demolished for a parking lot in 1929. Wright knew the building was due for demolition as early as 1925, FLW to Wijdeveld, 7 January 1925, in Pfeiffer (1982), p. 57. See also Neutra's photograph of the tiles in Sweeney (1994), p. 223.

39 On the first proposal see Alofsin (1993), pp. 97, 350,n67; Ron McCrea, *Building Taliesin* (Madison: Wisconsin Historical Society, 2012), pp. 65-166; on the Women's Building as built see Hamilton (1989), pp. 42-45; Storrer (1993), p. 168. See also the valuable John O. Holzhueter, "Cudworth Beye, FLW, and the Yahara River Boathouse, 1905", *WHistory*,

72(Spring 1989), pp. 162-198. Holzhueter convincingly argues that the project must be called the Beye Boathouse for the University of Wisconsin, that is by the man who commissioned Wright.

40 Rebori (1927), p. 453; Wright (1932), pp. 247-248.

41 Wright (1925), p. 59.

42 Wright (1932), p. 236, 245. On Freeman house see Dunham (1994), p. 59; and comments of owner "Samuel Freeman", in Meehan (1991), pp. 60-64; and so forth.

43 Wright (1925), p. 57.

44 Wright (1932), pp. 236-237, his emphasis, an amplification of Wright (1928), pp. 98-104.

45 Van Bergen had worked for Walter Griffin in 1907 or 1908 and otherwise in and about Chicago before professional independence. His Sears designs (Aurora and Carlton, nos. 3000 and 3002) were based on two houses completed earlier for his own clients. Aladdin's production slowed each year after 1928 but did not cease operations until 1987. Various web sources January-March 2012.

46 Wright (1932), p. 245; cf. Patterson (1994), pp. 4-5, 136-139.

47 Pfeiffer (24-36), San Marcos, pp. 56-3, details of blocks pp. 60-66, Biltmore, pp. 40-45.

48 As quoted in Onderdonk (1928), p. 164.

49 Tsuchiura, who stayed in Wright's office 1923 into 1925 before returning to Japan, is quoted in Alofsin (1993), p. 297; and Patterson (1994), pp. 149-150.

50 Levine (1996), p. 152.

6

Historians' Fiction

> The urge of nationalism ... encouraged Americans to rummage through their own prehistory so that they could equal or even possibly "one-up" Europe. Their own myth of an ancient past began to surface with the revelation of the pre-Columbian civilizations of Central America and Mexico, of the mysterious world of the Mound Builders of the Mississippi valley, and of the romantic prehistoric ruins of the American Southwest.
> — David Gebhard, 1999

> Midway Gardens: "Chicago marveled, acclaimed, approved.... It was 'Egyptian' to many, Maya to some, Japanese to others. Strange to all".
> — Frank Lloyd Wright, 1932

On many occasions Wright publicly chastised fellow architects and their clients for eclectically, imitatively, and occasionally slavishly, redesigning historical buildings: his and the modernists bugaboo about "historical presidents". The huge U.S. Supreme Court building by architect Cass Gilbert, finished 1935 as a white Roman temple, comes to mind as one of the most disappointing products of beaux arts fancy. Wright contended that to imitate the architecture of old senile *or* new foreign cultures was not modern, professionally degrading, even un-American. That is a matter of considerable record. He referred to those colleagues who borrowed not only Greco-Roman precedents, or kindred derivations born in the European Renaissance and reborn in the *classique* of nineteenth century France, but those Gothic and all other *ancient* Mediterranean Western styles. Yet, he often spoke of the enlightenment obtained from a knowledge of the forces effecting primitive or vernacular or indigenous buildings. He rejected the adulterating effect, as had Rousseau, of those old, tired, elitist and too sophisticated cultures that were trapped inside Tradition. The New World must free itself of that strangulating entanglement. Precedents were in universal fundamentals as expressed naturally by all preindustrial civilizations.

Wright's constant search was for a modern and an American architecture. Not the modern that some art historians described as beginning in Renaissance Italy around 1500 but, as Viollet-le-Duc had dreamed, one contemporary with Wright's generation. Such a search was the call of a small number of post-Civil War architects; Louis Sullivan their most eloquent and fervent, if tragic, spokesman. Wright took on the crusade around 1900 as if anointed the leader. Vociferously he argued the point in the late 1920s when he feared a new European hegemony, the new one spurred on by a kind of lazy objectivism, and into the 1930s when its dominance was a feverish fact.[1] He believed only his architecture was an antidote, curative, of and for the twentieth-century Americas. Therefore his designs of the 1930s were right for a Depression-gripped America.

To discover the why of the how and not the what—i.e. the principles—directing indigenous builders was the acceptable theoretical position. Thus in the late 1890s Wright sought to discover what Anthony Alofsin has referred to as the "primal origins of architecture". The investigation, however, was not of first (prime) civilizations but of those highly sophisticated: Maya, Aztec, Navajo, and Pueblo, for instance. Others such as Hindu or Ming or Javan were avoided in favor of those in New World geographic proximity, the pre-Columbians.[2]

Japan was exceptional and very popular. He enjoyed its traditional craft, not in total but aspects of its two dimensional designs. The aesthetic subtlety of its architecture, the Shinto attitude to nature and the unaffected application of local construction materials was complete, theoretically substantive. Buildings seemed to float in a surround of dark green. With equal vigor and still in search of a personal theoretical fodder, he studied meso-American architectural sculpture and indigenous Southwest American buildings. They sat square and solid on the earth. Each culture responded to the Nature given them—what else was there? Wright often inserted personalized adaptions of their decorative elements into a building's fabric not as an act of mimicry, but to visually emphasize the theoretical position.

The natural and organic use of old and new materials as argued by Wright's favorite theoretician, the French architect Viollet-le-Duc, was succeeded by the ever-appealing Arts and Crafts Movement. It initially looked at a laborer's craftsmanship in medieval England. To those in the movement, "natural" meant materials were left as close to their original organic state as possible, not disguised or smothered. (An example is a small detail: lead-based high gloss white enamel coating of wood window trim was incorrect. A person should see and feel wood in a native condition yet oiled or stained for preservation.) Son Lloyd recalled that the latest editions of the magazine *The Studio*, its London and New York editions, the Chicago arts and crafts magazines *House Beautiful* and *The Craftsman* out of New York state, were always on hand in father's office.[3] Their household products, too often naively designed and ill-proportioned, were similarly "natural"—if not as refined and elegant—as traditional Japanese decorative design and building construction.

When a teenager, his revered uncle, the Reverend Jenkin L. Jones, persuaded Wright that Nature was the provider and great designer. Emersonian in content, this was a transcendental truth that suited the architect throughout life. Euronorth Americans believed that native Indians were finely tuned to nature.[4] Wright's experiences with Indians began when his grandparents claimed Indian lands so they could farm. His youthful summer encounters, and later his mature artistic thoughts, centered on the plains people, hunters and gatherers of the upper mid-west. They had no need for permanent buildings but their artefacts were well known and meanly borrowed. From 1890 into the 1930s some of Wright's decorative designs, wall paintings, and architectural sculpture were obviously based on Indian designs; for instance, a large wall painting in the play room where Wrights' wife, Catherine, ran a neighborhood kindergarten in the 1890s to around 1906.

Southwest Native American architecture was permanent, extant, and close to a true primal architecture where first principles, social and constructional, were clear. There, old and new sand colored adobes made of grit and soil, or stone cut from the soil, were resolute geometric architectural forms firm on the earth, plain in sun's bright light and under sky's blue. Therefore a pre-Columbian influence was not apparent in Wright's *buildings* until 1915. That was after personal experiences during travels through the Southwest and on to San Diego. There he revived friendship with a former colleague in the Adler and Sullivan office, architect Irving Gill.

SOCIETAL QUIRKS

One of the design disciplines, architecture has a peculiarly intricate yet expandable design process, one increasingly complex since the English industrial revolution began in the eighteenth century. That process includes the client's program, essentially why and what they want built, and a multitude of intellectual, physical, psychological, and highly technical infiltration's, all coordinated by the architect. To construct a building requires the artefact of documents that explain what is to be constructed, and thereafter the occupation of many people and diverse mechanics over a period of time. Again the coordinating supervisor is the architect. As a design form, a building's physical appearance should, to most practitioners (post-Vitruvius but pre-Venturi?), should expose—or at least clearly express—the way it is made. This confirms it as a design product that reaches culmination in functional habitation. To paraphrase architect and teacher Louis I. Kahn, the making of a building should be experientially clear to its users. To quote Wright: "the element we call structure is primarily the pure form, as arranged or fashioned and group[ed] to 'build' the Idea; an idea, which must always persuade us [the user] of its reasonableness".[5] That was one theoretical tradition Wright favored, at least up to the 1930s, and one purpose of his investigations into native American buildings, or at least into their external appearance.

Within the architectural design process two basic contrivances are employed to explain design to designer, to client and to builder. The main—or principle or ground—floor plan sets the idea, organizes function, and contains the implications of gravitational resistance to constructing a third dimension that will result in the physical definition of spaces. Wright correctly referred to "ground plans" as the "actual project[ion] of a carefully considered whole".[6] The main floor plan, therefore, defines architecturally the idea and identifies to the designer, builder, and observer, a building's various horizontal and vertical systems and spaces, and the enclosing and embracing forms. Elevation, section, perspective drawings and a scale model, then picture the third dimension of the future building. Today we use computer generated pictorial images that can include sequentially changing perspectives. Yet, like a drawing they are a two dimensional visual presentation, *not* virtual space but descriptive.

Unless sympathetically gifted, those historians and observers who have not experienced the complexities of designing a building, or the highly involved construction processes, analyze architecture at a disadvantage. John Dewey has convinced us that art, and surely the design discipline of architecture, is an experiential phenomenon. The lack of design experience is most noticeable when the untrained observer attempts to evaluate complicated, post-1850 buildings. An architect can usually spot such a person's writing. Wright for example, and in step with most designers and artists, was not disappointed but generally disgusted with evaluations by posing intellectuals and academics and by the untrained and inexperienced who, to him, were unknowledgeable and in that sense ignorant.[7]

It should be clear that the indigenous and vernacular architecture of each culture can be identified by floor plan arrangement, construction materials as found locally, and composition of the three dimensional enclosure. To this is included—or added—the esoteric refinements we call ornament. All this was well understood by central European modernists and their American followers in the 1920s and 1930s. Without ornament, with standardized non-local but universally available synthetic materials (like glass, steel, concrete), with "functional" plans expressed (a proposition seldom realized), and with aesthetically non-stylistic enclosures devoid of societal and ethnic quirks, political and cultural identifiers were eliminated. The idiosyncratic architecture of Armenia or Japan or Mexico or India or Spain was meant to disappear, supplanted by neutral boxes defined by a hard-edged, abstract machine aesthetic. (Immediately, the buildings by Morrill and later the German/American architect Mies van der Rohe come to mind.) When rationalized it became known quite correctly as the International Style. Wright vigorously opposed both its political pretensions and, he argued, its devaluation of personal and cultural identity.[8]

WORDS AND EDIFICATION

> *I believe that out of the past comes the best of the present and that out of the present comes the future.*
> — Frank Lloyd Wright, 1932

It is easy to misunderstand what Wright meant by "out of the past". Rest assured he was not talking about architectural styles. Those who analyzed Wright's architecture, with few exceptions, were/are not architects or trained in design.

Again with few exceptions, they generally employed three analytical methods. They either took what Wright said about himself and his works and repeated it, and this prejudiced understanding. Or they looked with the less narrow vision of a traditional art historian. In this case "traditional" infers the methods inherited from nineteenth century purposes, religious or political, and conditional methodologies. Neither has proven adequate, separately or in combination. The first and second have been identified and tested by only a few who employed the third method, that is to look for and study evidence with a bias toward architecture and the architect. Those who study Wright with this method include biographers Robert Twombly, Meyrle Secrest, the lurid prose of Brendan Gill, and historians Thomas Hines, Anthony Alfosin, Jonathan Lipman, Kathryn Smith, Paul Kruty, John O. Holzheuter, Otto A. Graf, Joseph M. Siry, Kevin Nute, Donald Leslie Johnson, and Paul E. Sprague along with his select colleagues in Madison.[9] Also included are most of the authors assembled by Robert McCarter for his collection of essays *On and By Wright*: John Sergeant, David Van Zanten, Gwendolyn Wright, Patrick Pinnell, Richard MacCormac, and Smith, Graf and Lipman.[10] The style of the first method can be rejected outright. The problematic style of the second will quickly become obvious.

An article of 1953 by art historian Dimitri Tselos, who found an "evident dependence", he said, "of Wright upon Mayan architecture", initiated what became a shower of commentary by people who, like Tselos, found "exotic" borrowings in Wright's architecture.[11] Nearly all of those findings were about small parts of a building, things decorative, occasionally an isolated surface or shape. And most suggestions were dubious visual analogies empty of knowledge about the complexities of context, architectonics, and building construction. Usually they were about things "tempting" or "suggestive" or with certain "parallels", or "echoes" or "synonyms". On the basis of these an entire building was assigned a type or style. However, readers were not informed about the overall architectonic consistency of the suggested sources or how they were integrated, or not, with a particular and complete Wright building. They invariably ignored architectural fundamentals such as the floor plans of the echoed buildings or of Wright's. They did not compare architectonics or construction techniques or peculiar inclusions or conditions that orchestrated Wright's aesthetic responses.[12] Conventionally, the comments were at best provocatively judgmental and aesthetical impressions. At worst and to repeat, they were mere visual analogies: the great majority incorrect.

There is no doubt Wright adapted minor elements but only now and then, here and there. As mentioned above, his reason for inclusion was to amplify a theoretical position by a visually identifiable referent. We shall soon cite some of these. Serious problems arise, however, when findings and opinions extended beyond minor detail, yet fail to include the entire building, Wright's or those previous others.

In the case of his Southern California houses of 1919-1925, predecessors usually selected by commentators were buildings created by Central America's pre-Columbian societies. The Aline Barnsdall house, a related proposed theater on Hollywood's Olive Hill, and the textile block houses, were said to be "definitely suggestive" (a positive perhaps) of certain buildings by meso-American cultures.

FIG. 51.—Governor's House at Uxmal.

FIG. 52.—Ground plan of Governor's House at Uxmal.

Fig. 6.1 Exterior view and ground level plan of the so-called "Governor's House" at the Mayan temple site of Uxmal in Yucatan, Mexico

The Millard and Ennis houses were on other occasions said to be, yes, Aztec.[13] Tselos found a "parallel" with the Millard house in "the inner court of a Zapotec palace in Mitla, Mexico". Moreover he said "the same palace" should be compared to the "plain and decorated stoa of the Imperial Hotel".[14] And so forth. Fig. 6.1 shows a typical Mayan temple floor plan and an exterior view that is typical of pictorial presentations published prior to 1910. Only public, royal or religious buildings were constructed of stone and are extant. Houses and such just disappeared, absorbed by the jungles of which they were made.

While the North American Indian "was being fashioned into a symbol of continental uniqueness", as historian Henry Schmidt has nicely and politely summed, there was a deepening interest in all aspects of Mexico including its apparently simple life style. But most attractive was its monumental architecture at pre-Columbian archaeological sites that had remained unknown to transgressors until 1863. In search of the "exotic qualities" of the indigenous people and of the impact of Hispanic conquest, all sorts of North Americans traveled south during the last four decades of the nineteenth century. But the Mexican Revolution began in 1910 and thereafter travel was discouraged, often dangerous. When the killing stopped in 1920 and stability was forecast, tourism as well as serious archeological investigations regained momentum.[15] By then Wright was desperately trying to stabilize his post-Tokyo private life and to restore a professional career in the United States. While so engaged he maintained an ongoing and successful defense of his unique architectural products pre1922, especially in European publications. He vigorously promoted Barnsdall's house as a critical transition to professional revival.

In any event, there is little to be gained by recalling here those many guesses as to precedent and borrowing that had begun as early as about 1914. Anthony Alofsin within an art-historical construct, succinctly outlined many of those opinions and provided reasoned arguments as to the variety and *universality* of possible sources, two and three dimensional. They included themes found in Greco-Roman, Mayan, and Viennese Secession.[16] Alofsin's concern was real enough because, after Tselos's article, even architectural historians were offering opinions, more sentiments than not. Yet the character of guesses can be measured by mention of just a few: "The Barnsdall House invokes Mayan temples . . . Its platform is compressed [as a] Mayan geometricized hill mass". And: the Kaufmann house Falling Water of 1935-37 "shows that Wright had finally found how to use the Mayan pyramid".[17] Or another: Millard's house "brings to mind the Temple of the Sun at Palenque, not as a prototype or model, as in Hollyhock House, but as a resonant after-image".[18] And another: "Wright concrete 'knit-block [sic] houses . . . [seem] to begin Mayan and end Islamic".[19] Indeed, observers from all walks thought Midway Gardens was a Mayan edifice, maybe Aztec, perhaps Chinese, Egyptian?, surely not Gothic. More recently we were informed that there was a direct match with the Maya, that the Barnsdall house "is the plaster model of a Usumacinta temple" on the Yucatan peninsula.[20] And so forth.

Those declarations of reproductions, replicas or imitations by the learned have been forthcoming for decades. So it was inevitable that non-historians, journalists, social biographers, even archaeologists, have repeated them willy-nilly. Social recorder and journalist Brendan Gill, for example, once described the Barnsdall house as "an exotic Mayan fortress".[21] Cultural historian James Oles told us that Hollyhock "incorporated massive walls and repetitive surface decoration derived from Classic Maya models". There is no surface decoration, the walls hardly massive, and the models not identified.[22] In 2004 a columnist referred to the Tokyo Hotel "with its stone and brickwork reminiscent of some Aztec or Mayan structure".[23] Author and musician David Hertz believed that the German warehouse metaphorically blends "the Mayan design with the Shinto shrine—where grain was stored on a raised floor".[24] And so forth.

The simple acts of collating research and of visual analysis has produced a confident, more rational analysis as follows. Hollyhock is *not* made of plaster and is not a model. It was designed to be a house not a religious edifice or royal or governmental building. Beyond that, and true for *all* of Wright's buildings, the *floor plans* are not Mayan or that of another meso-American, Central American or North American aboriginal culture. The physical *enclosure*, the construction *materials*, the *scale* and *proportional* systems (two and three dimensional), and the *colors* are not Mayan or other derivation. Barnsdall's *interior* spaces, their surfaces and their *decoration* are (for the most part) typically Wrightian of previous decades, not Central American. (Pre-Columbian house interiors were adobe or lime wash or exposed grass. Public buildings were stone or, where adobe, coated with a lime wash or some form of stucco often painted in bright colors.) To paraphrase Wright: the Barnsdall Idea, as he emphasized, and its appearance were not Mayan; rather they were of its own geographic region. Further, he *intended* the—or should that be his—building be Wrightian, inspired by "primitive" impulses and natural processes.

Moreover, it is difficult to believe that for semi-arid earthquake prone Los Angeles, Wright would imitatively borrow the ceremonial stone architecture of the hilly, deep green tall damp and steamy jungles of Central America. Wright was obviously not interested in creating a modern (new) Mexican architecture. However, we shall learn that the *massing* is recollective not of the Maya or Aztec but of Navajo, Walpi and Pueblo.

Other than a couple of decorative things and figurative sculptures, two isolated Hollyhock elements have been referred to as Mayan. One was architectural: the slightly canted or inclined attic (in classical parlance). The other was a row of sculptural abstract hollyhocks standing on an exterior projecting string course (or band) at door head height. The inclined attic has been described this way: "The sham wood and stucco walls and canted roofs imitate the shapes of Maya temples in an unusually direct, historicizing way".[25] But the wood, stucco and walls are not sham. Obviously the roofs are not canted; they lay horizontally behind the canted attic that practically is a series of short walls acting as a handrail or balustrade around the principle flat roofs. They are framed in timber or ordinary concrete block. The wood, the walls, the attic, the roofs have nothing to do with historical eclecticism. The cement stucco exterior surface will be discussed later herein.

Three preliminary elevational drawings of Hollyhock prepared by Schindler are known. One recalls Wright's northern prairie designs with no attic and a persistently level gutter line. The second had a string course in relief and is akin to later studies for the aesthetically frenetic and unrealized desert "compound" for A.M. Johnson, or to the overly complex proposal for a new Barnsdall house for Beverly Hills, both designed in parallel during 1923.[26] Preliminary drawings of the third and final design for Hollyhock also confirm that the exterior incline of the attic was determined very late in the design process, almost as an afterthought: not even anecdote tells us why.

The other element was decorative. Apparently in response to Barnsdall's request, sculptured precast concrete hollyhocks were added after the design basics had been complete: again an afterthought?[27] In a clever design, abstracted hollyhocks sit upright against stuccoed walls and on a projecting string course. Schindler described it correctly as a "lintel course" because the bottom line of the course was set at door and window head height. And the word "course" is in this instance used freely. Coursing normally refers, of course, to the horizontal lines or courses in masonry construction.

Also, the precast hollyhocks in style, detail, and appearance, singly or collectively, cannot be found in Mayan sculpture or architecture. Hollyhocks are not indigenous to the Southwest or meso-America and require considerable water. As previously mentioned herein, their design was taken from the conventionalized floral sculpture that formed the colonette capitals on Unity Temple, but inverted and called Hollyhock. And they were applied only on the exterior of the living room, of the upper floor bedrooms at the rear, and as capitals on courtyard posts. During the design process, on elevation drawings of the principle volume only, they replaced an earlier design for a line of squarish "tiles", as Schindler called them, of a style similar to a few in terra cotta found at the Tokyo Hotel.[28] Similar but flatter tiles were used on parts of Residence B,[29] in a manner reminiscent of a few of Sullivan's buildings and tombs on which the young Wright worked. Other finials (not uncommon in his buildings of the teen years, notably at Midway Gardens and the Tokyo hotel) and sculptural elements for the Barnsdall house were also precast by a "concrete stone" company, or in "art stone" in historian Smith's terminology. A couple of those elements might look rather Mayan to some viewers. Ordinary concrete blocks were used for attic walls in some places and stuccoed.

Further and as an interesting historical connection, the principle Barnsdall elevations are predated by the Griffins' designs for a chemistry building and a library for the University of New Mexico in Albuquerque. Walter was invited to the university in 1913. On a tour of nearby pueblos he was thrilled to admiration for the "handsome, finished, beautiful, compact, sanitary, convenient cities", he said, with their natural "homogeneity, unity and simplicity".[30] The university believed a modern interpretation of those adobe pueblos would represent an "authentic American" architecture, in contrast to the then popular "University Gothic" passed on from New England.[31] The university's president was clear: Griffin would make UNM "as distinctive as Leland Stanford [University] is", differing only "in that it would be based [on] the lines of the Indian pueblo style combined with the features of the so-called 'Mission Architecture'".[32]

Walter and Marion responded. Drawings of 1915 of the exterior elevations for the chemistry building, Fig. 6.2, show a pronounced string course in high relief at lintel level (therefore a lintel course) and a tall attic, actually the exterior wall of the second floor. It was to have an internal clerestory lighting and space for fume exhausts, shown in the section drawing. The elevations are remarkably similar in character to Hollyhock, but not the ornamental refinements. The floor plan and fenestration are of course different. Walter referred to his buildings as a "pueblo type", not a style. The design was not realized but the university persisted with buildings in the then fashionable Pueblo Style. Wright almost certainly learned of the proposed New Mexico buildings from Frank Barry Byrne, his sycophant, a former Wright and Griffin employee, acted as partner-in-charge of American commissions while the Griffins were in Australia, but only during the years 1914 into 1915. Byrne modified the Griffins' design, perhaps to meet university concerns, in 1916, the year Byrne took full control of the project. The building was complete in 1917 to Byrne's design and is heritage listed.[33] String courses (but not lintel courses) of more modest expression were also found on Pueblo buildings and on Spanish colonial revival styles.

So, if looking for inspirational sources for Hollyhock there might be many. Yet, it is clear that for the floor plans Schindler and Wright drew on their own experiences and for the massing and elevational character they looked to the indigenous American southwest.

To clarify differences between Wright's and Irving Gill's modernism and the eclectic so-called architectural revivals so very popular in the Southwest, the example presented here is the second Owls' Club in Tucson, Arizona, of 1903 by architect Henry C. Trost. Fig. 6.3. For a short period he was a follower of Sullivan and Wright in the Arizona, New Mexico and West Texas area, The Club has a *plateresque* tablet with Sullivanesque stucco designs in relief and a little Pueblo mixed with a lot of Spanish colonial and plenty of stucco. Trost added an oriel window with a Sullivanesque ornamental surround (also stucco in relief) and a string course with

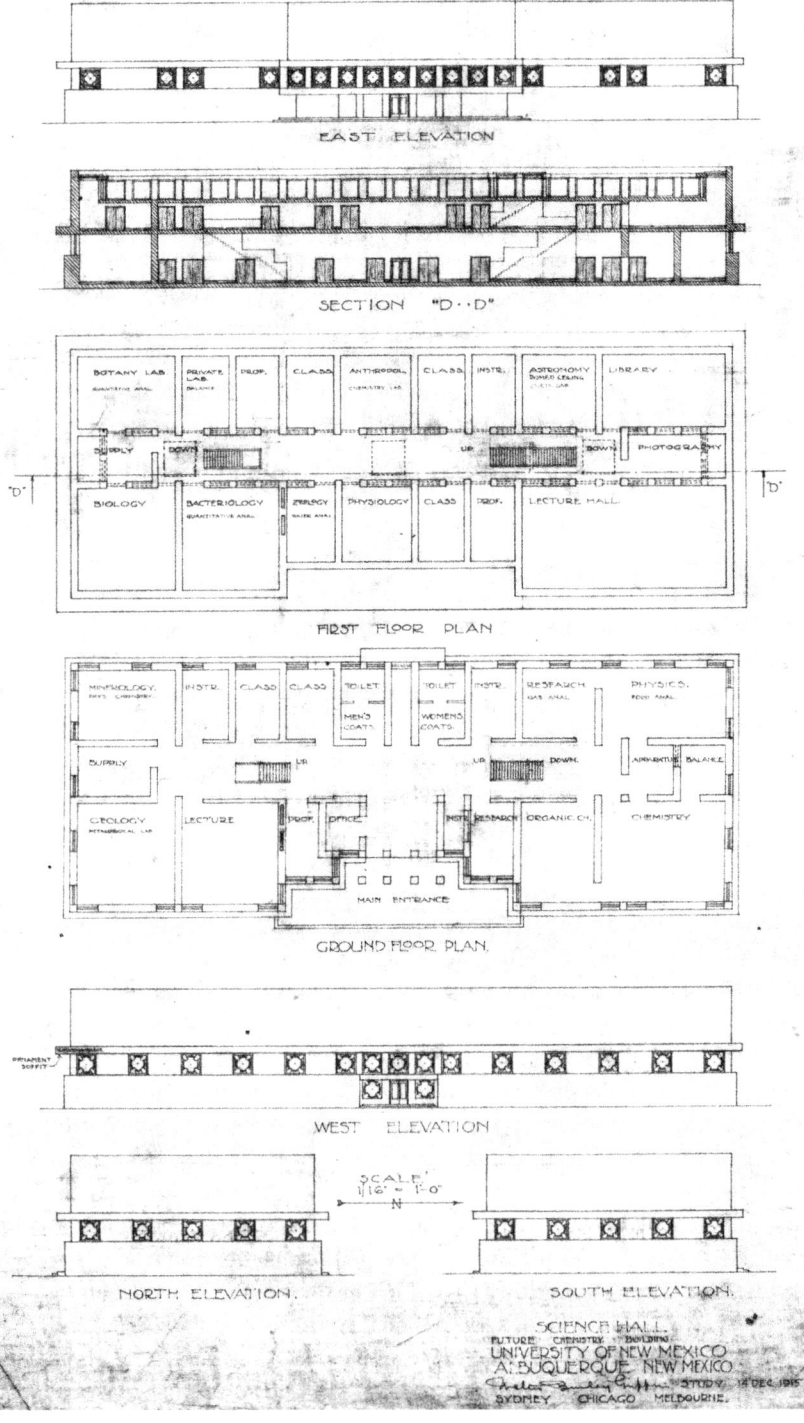

Fig. 6.2   Project for a "Science Hall future chemistry building" at the University of New Mexico, 1915, Walter and Marion Griffin, Architects

Fig. 6.3 Owls' Club in Tucson, Arizona, Henry Trost, Architect, of ca. 1903

attic above that contains protruding imitation Indian or Mexican *canales*, some of which actually work.[34] The appearance of the original interiors are unknown.

A more masculine building was the Franciscan Hotel designed and constructed during 1923 and 1924 in Albuquerque, New Mexico, by the firm of Trost and Trost (Henry and brother Gustave) then out of El Paso, Texas. The "sun-baked concrete walls", as then described, and general character of the exterior and interior was understood to be derived from ancient pueblo, colonial and mission Spanish, and territorial precedents. Yet its bulky and massive forms stripped of ornamental impedimenta, fake *canales* and scuppers, and a nod to modern office-building verticality gave the building a certain grave gravity.[35]

As a series of preliminary drawings for Hollyhock reveal, Wright was ambivalent about the attic: to attic or not to attic, straight or canted. But its slope was incidental to the concept.[36] Once the decision was made to use the flat roofs as sun decks (off Barnsdall's bedroom) and for soirees and drama productions, soon to be discussed, hand rails became essential. Moreover, as a basic design problem a tall straight sided attic, if viewed from below at middle distance, will appear to flare out. This is true of any relatively tall rectilinear mass above eye level regardless of converging perspective. On Barnsdall's Beverly Hills house project, because the materials below the string course were to be highly textured blocks, and above them plain stucco, the visual effect is slightly more noticeable. Even in drawings, the attic on the Griffin chemistry building appear to flare. Since the Ennis house sits high on a hill and is seen

Fig. 6.4 Warren Hickox house in Kankakee, Illinois, 1899-1900

most often from positions well below, as is Hollyhock, Wright *may* have employed stepped and canted walls to minimize the apparent flair. The second floor exterior walls of Lloyd Wright's Bollman house were constructed of stucco on wood studs and vertical. In construction drawings of 1922 they were designed with a cant.

A couple of other examples of errant claims of precedent reveal the dimensions of opinion. The Warren Hickox house in Kankakee, Illinois, of 1900, Fig. 6.4, and a few other houses by Wright of that period, have been described as looking very much like traditional Japanese architecture. This does not hold up to scrutiny. In the last half of the nineteenth century plaster and batten walls from Bavaria and Tudor half-timber, for instance, were widely used as were similar effects of English and American Arts and Crafts. Elsewhere it has been shown in detail that all other aspects (floor plans, materials, proportion, scale, etc.) of this and other of Wright's pre-1917 house designs, were either typical of their time or learned principally from Joseph Silsbee, less from Alder and Sullivan, and well *before* 1893 when a Japanese pavilion was built on grounds of the Chicago Worlds Colombian Exposition.[37]

It is misleading to say those domestic buildings were derived from Japanese architecture because of a few dark-stained timbers with a plaster infill or a roof shape or wide over hang (both universal) but ignore, as explained above, the plans, materials, proportions, and all else. However, Wright's debt to Japanese art is only slightly noticeable on a few buildings by their general simplicity of line and mass, and by the use of square proportions. More particularly it is seen in his architectural presentation drawings before 1914, and they may have swayed inattentive observers. The foreground foliage in Fig. 6.4, for example, nearly copy oriental drawings. And there is the entire perspective drawing of the first Millard house of 1905.

Another example. Following Tselos' lead, in the 1960s it was stated emphatically that the Albert Dell German Warehouse, Fig. 6.5, in Richland Center, Wisconsin, is,

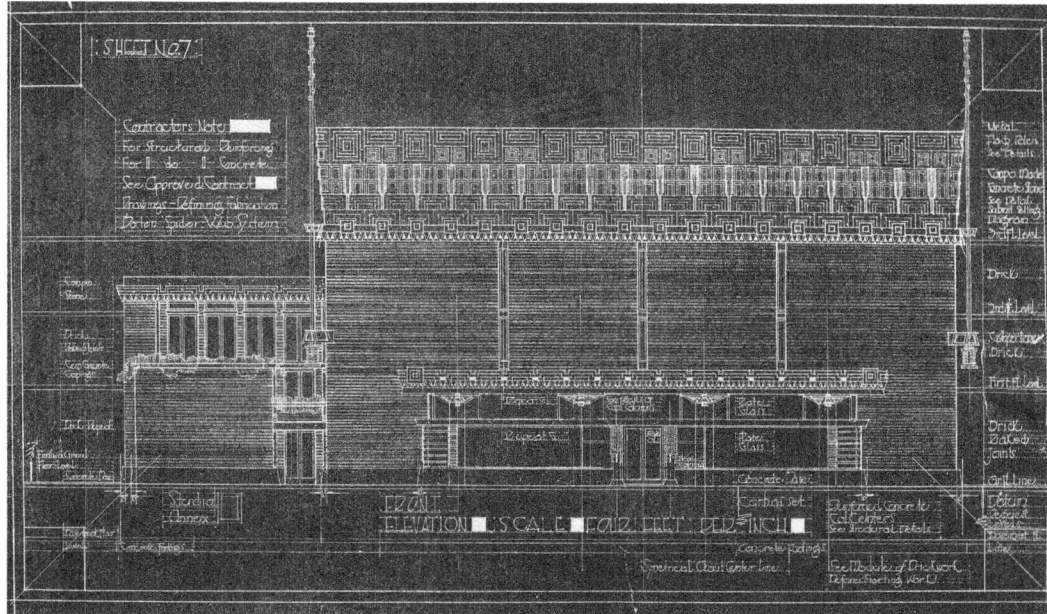

Fig. 6.5 A.D. German warehouse in Richland Center, Wisconsin, 1915-20

yes, it "is the Temple of the Two Lintels". Or perhaps the Chichan-Chob, both at the Maya site of Chichén Itzà on the Yucatan Peninsula.[38] Then in the 1990s there was the expected reiteration and amplification:

> Here, for the first time, the overall configuration of Wright's building is pre-Columbian, specifically recalling both the Temple of the Two Lintels and the Red House (also known as Chichan-Chob) Chichèn Itze.[39]

The facts are otherwise. The warehouse in floor plans, elevational parti, construction details, and structural technique was determined by Wright's experiences with Adler and Sullivan, in particular the brick clad concrete structure of the Chicago Cold Storage Exchange Warehouse of 1890, and in 1892 and 1893 the stage house exterior for the Seattle Opera House project: each also had vertical slit windows exactly like Wright's warehouse.

The German building's exterior, designed in 1915 immediately after Wright's visit to San Diego, has a classical Greco-Roman presentation of base, shaft, and termination. Yet the building's proportions are *not* classical Mediterranean *or* meso-American. In elevation we see the base is a thin line of concrete; the shaft is brick set in Flemish bond; the tall complex quasi-attic (a much smaller but very similar attic was on Sullivan's Cold Storage building) acts compositionally as an encircling terminus, perhaps a frieze; the exterior walls are not load bearing and their bricks are yellow stretchers and red headers. Yet the building's proportions are not classical Mediterranean or classical meso-American. The individual precast concrete elements of the bold attic are derivative of—but unlike in detail—meso-American architectural stone sculpture as found on facades. But clearly they are *not* from specific sculptural antecedents and therefore *not* from a specific building. Rather, taking a cue from Sullivan, Wright applied a series of precast concrete tiles with relief patterns based on a square that were similar to European and other diverse precedents that could also be found on Maya buildings. Each row is set out from the lower so the attic actually flares outward; in white concrete against yellow/red brick it dominates. Wright used those precast concrete designs again at the Tokyo hotel and for two of the block designs on the Storer house. The top row of blocks with the large square pattern act as a balustrade at roof level. Below them at the fourth floor and within the block pattern there are slit windows. Concrete coarse as a ledger to the attic is also formed of a dove tail molding but modified. Two concrete corner braces for a vertical ornament or for flag poles on the street facade were taken from the Maya or Ptolemaic Egyptian as another referent to what Wright referred to as the primitive.

Therefore the German floor plans and other architectonic parts leading to or composing the whole were *not* based on Mayan buildings. As well, the ground floor entry and its side windows as designed early on (and not as built) were derived from a variety of commercial buildings in America. On sheet seven of the construction drawings there is a note:

> *For Structural Reinforcing*
> *For do Concrete*
> *See Approved Contract*
> *Drawings Defining Fabrication*
> *Datori Spider Web System*[40]

The exterior concrete columns with square mushroom capitals ornamented with one dove tail molding pattern in relief were also found in the building's interior and unadorned. The facia of the original entrance canopy was in one drawing to be a relief based on American Plains Indian abstractions of winged eagles or (less likely) of Egyptian depictions of their mighty winged sun god. Those few "exotic" elements were not architectonic but ornamental. Quite simply the "overall configuration" of the building is a top-heavy box with an exterior of brick and concrete tiles as ornament.[41] Why so much exposed concrete? Quite simply, one of German's products as a retailer was bags of Chicago-AA Portland Cement.

Tselos did make a contribution to our understanding of the attic's ornamental sculpture when he said, "The abstract and zoomorphic ornament . . . carved on stones imbedded in the [external Mayan wall] fabric . . ., Wright substituted a truly abstract geometric ornament impressed into [concrete]".[42] The word "substituted" is a strange qualification to an observation that otherwise relieves any accusation of mimicry. Therefore, the *general character* of one part of the terminal ornament is partly Mayan, but the building is otherwise most certainly not Mayan. German stopped construction of the warehouse in 1921, the building never completed.

There are human-headed winged eagles (hiding a storage attic) boldly positioned on a stone-like frieze above the street front window lintels of the Bogk house, a concrete and brick structure in Milwaukee, Wisconsin, built from 1916

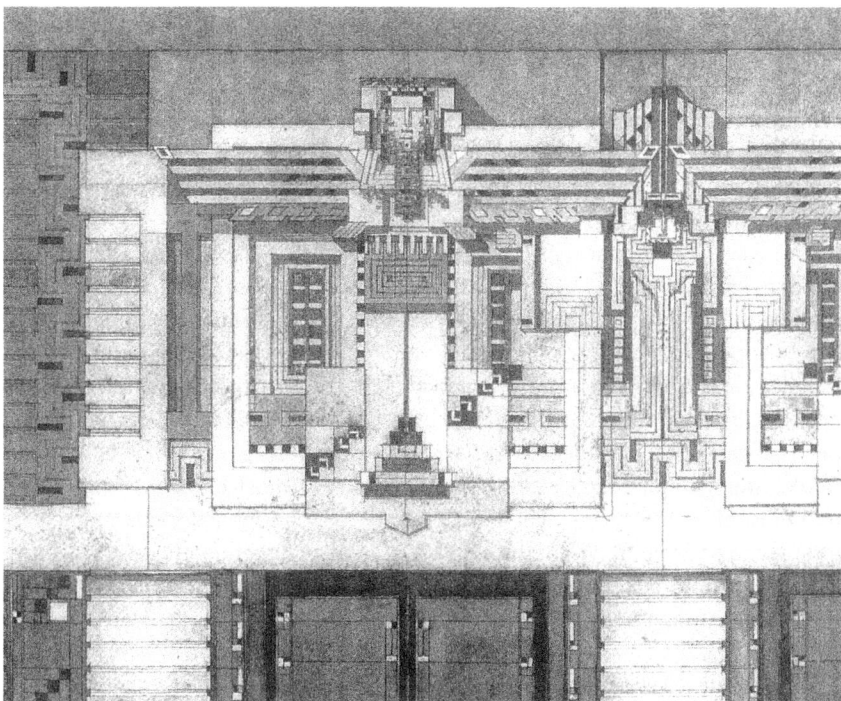

Fig. 6.6 Frederick C. Bogk house in Milwaukee, Wisconsin, 1916

into 1920. Fig. 6.6 is the original design but when built it was less complicated. But again the eagle ornament was an afterthought.[43] Although often referred to as pre-Columbian, the design is an abstraction of American plains and Southwest Indians' own abstractions, but here in precast concrete or terra cotta in high relief.[44] The house is Wright School of the period 1902 to 1914, inside and out.

From ca.1895 to around 1915 Wright and author Hamlin Garland were close friends. Garland wrote about plains Indians, Euroamericans in the newly ruralized Mississippi Valley, and the spoiling of the unspoiled West. Both men were besotted by Indian themes, even costuming their families and friends for picnic parties.[45] It is not surprising that the winged eagle theme was abstracted even further and placed on the Ennis facade outside the dining area. Son Lloyd used a similar reduced abstraction on the front facade of the Samuel-Navarro house in Hollywood of 1926 to 1928. And images of the wing theme appeared at a coupled of locations on exterior faces of the Imperial Hotel.[46]

As an aside, we note that the historian's rush to find visual precedents for ornament (except for historian Neil Levine), the obvious has escaped notice, so easily at hand: the line and geometric patterns created by Indian basket weavers and potters. But where do the off-hand visual analogies stop; Tselos and followers teased readers with dozens. In fact comments and analyses made since 1908, some by reputable observers, attributed precedents for and sources of Wright's buildings to Greek or Roman antiquity, to the European Renaissance and Baroque but not Gothic, to "the Vatican", Malta, English Arts and Crafts, the bungalow craze, ancient Egypt, Syria and China, the Pueblo, Navajo and plains Indians, Maya, Aztec, Tlaloc (but not Olmec or Inca), Japan of course, European abstract impressionism, Austrian *secession*, and who or where or what else? Wright was probably amused to read those incautious assertions, and found no need to rebut them directly simply because, as superficial as they might have been (or are), they were *in part* true. But not for the reasons their authors did or did not offer.

From all this we learn that Wright would not explicitly imitate primitive or indigenous or foreign or antique architectural predecessors except as he might ornament a building. Historian Joseph Siry has put an intriguing thought. Wright "transcended such [historical] models to signify [their] cultural present". More rationally and practically, and as Wright would have had it, they signify a theoretical position. In other words, their reference on Wright's buildings was, Siry said, "an original synthesis, intended to be a contemporary statement that condensed formal and symbolic ideas from multiple pasts".[47] It must be again emphasized that most often they were relief or figural or two dimensional designs. Many were introduced to a young Wright through his favorite architectural theorist, Viollet-le-Duc who, through his writings, tried to understand the exotic.[48] See Fig. 6.7.

Wright historian Anthony Alofsin believes that . . .

> *Wright believed that a "spell power" [Wright's words] existed in primary forms, which allowed the creation of an iconography that had cultural significance. An artist could take indigenous forms with received meanings and transform them to represent contemporary social values in what Wright called "emulation . . . [not] imitation" of tradition.*[49]

Thus, Wright could identify the ancient Greeks' own antiquity as "one starting point for the [Western] classical tradition". So, his architecture prior to the 1920s was at various times formalized by axial symmetry, classicistic plan parti, and massing even in the wilds of Montana's Bitter Root Valley.[50] Because formalization was a universal derivative it could also relate, as he put it in 1953, to "styles associated with a primitive ideal, such as the Egyptian, Mayan, and Japanese".[51] But to repeat; those were not primitive but highly sophisticated societies, their association with primitive ideals marginal except as tribal history. Perhaps Wright meant fundamental ideals rather than primitive.

Architectural critic Douglas Haskell, after discussions with Wright in 1928, resolved the messiness of attribution and precedent in an essay too easily forgotten by academics reading other academics. As editor of *Creative Art* the perceptive Haskell noticed that eclectic New York City architects McKim, Mead and White:

> *took "Italian Renaissance" and modeled it; [Wright, however,] ranged over the world's architecture and digested it. Not for nothing are some of his constructions related to Japanese, some to Maya, some to Egyptian art. He speaks, himself, with gratitude, of the lessons these and other civilizations teach.*[52]

Fig. 6.7 "Portion of Exterior of Nahuan Palace" published in *Viollet-le-Duc* (1876)

"Related" as associated through those common fundamentals and easily through an examination of ornament; and lessons, not tracings. Siry refers to this duality as a "kinship... of the past". One should note that as far as Wright was concerned, this was not a difficult intellectual tension as it emanated from a flexible mind. On reflection Wright's reduction was precise: "And yet altogether these architectures seem to acknowledge kinship to each other, whether Mayan, Egyptian, Dorian or Chinese".[53] Ada Louise Huxtable was unimpressed with architectural historians' indecisions, then spelled out her own:

> *In no established or identifiable style, the [Barnsdall] house bears the marks of many; the influences have been variously described as Aztec, Mayan, Egyptian, and Mexican pre-Columbian. If the look was archaic and vaguely Mesoamerican, the symmetrical plan and flanking wings still had a Beaux Arts formality.*

To Huxtable the Ennis house was like a baronial fortress, Millard's house had a "suggestive Hispano-Moresque aura": aura. And so forth.[54]

We can now add a note in summation. In the early years of his career Wright acknowledged that he strove for a certain picturesqueness. The thoughts of art historian John M. Jacobs Jr. are appropriate in that he does not worry about architectural Styles.

> *Besides being picturesque (and here Wright is the heir to a long Anglo-American tradition) his idiomatic forms often seem to be distilled from some secret, highly evolved hieroglyphic script, and hence to be the bearers of concealed representational or ideographic messages.... [A]rchitectural forms seem on the threshold of becoming a kind of sign language.*[55]

In 1953, in the first direct reference to followers of Tselos, Wright replied with his own emphasis: "Had I not loved and comprehended pre-Columbian architecture as the *primitive basis of world-architecture* [he emphasized], I could not now build as I build with understanding of all architecture".[56] (The *fundamental* basis, perhaps.) He was reiterating an explanation he made in 1930 to a Chicago audience:

> *Whenever architecture was great it was modern, and when ever architecture was modern human values were the only values preserved.... [T]he principles moving us to be modern now are those that moved the Frank and Goth, the Indian, the Maya and the Moor. They are the same principles that will move Atlantis re-created.... Principles are universal.*[57]

Those statements were a defense of commonness and reinforced the notion that all architecture is related by natural functional determinants and cultural necessities, and by irresistible societal conventions, all exalted in architecture only, Wright would proscribe, only when it respected the mother force, Nature. The specific example of pre-Columbia as raised by Tselos in 1953 presented Wright with an opportunity to make a relatively exacting explanation of the *theoretical* facts, the summation immediately above. Then in 1954 Wright offered a refocus, one that may have tempted Siry and Alofsin:

> *People often ask why I, a modern architect, have so many old things around [my house]. Why not? I, too, belong to tradition—back to the oldest American architecture ... and to the Japanese and others. All of them are brought into now.*[58]

All signifying continuity by a "cultural present". We are conditioned by historical evolution, therefore we study history to better know ourselves. What has been is now and always will be, to paraphrase Louis Kahn. History and the earth, what else is available to human life?

Again in the 1950s a rather frustrated Wright said:

> *To cut ambiguity short, there never was exterior influence upon my work, either foreign or native, other than that of Lieber Meister [Louis Sullivan], Dankmar Adler and John Roebling, Whitman and Emerson, and the great poets worldwide. My work is original not only in fact but in spiritual fibers. No practice by any European architect to this day has influence mine in the least.*
>
> *As for the Incas, the Mayans, even the Japanese—all were to me but splendid confirmation.*[59]

We were wrong when earlier we said the U.S. Supreme Court building is a Roman temple. Rather, it is a large and expensive office and court room box clothed by marble with a grossly scaled Roman temple front appended; all very whitish. In this respect it is not dissimilar to the Virginia State Capital building, Thomas Jefferson the architect in consultation with the French architect C.L. Clèrisseau in the 1780s. But philosophic and theoretical promptings for the two temple fronts are quite different: the older was part of a revived Greco-Roman classicism in all arts and letters. The newer was part of a devolution to a fancy for architectural styles. It was a continuation of nineteenth-century thought when architecture was ornament.

When in 1915 Henry Blackman Sell wrote a piece about Wright for the New York *International Studio* he correctly entitled the article "Interpretation, No Immitation: Work of Frank Lloyd Wright".[60]

Today it is quite incorrect, even absurd to claim an entire building (by Wright or anyone else) to be a copy of or imitative of another simply because one or two rather superficial or extraordinary exterior elements *look* something like a predecessor. Looks are deceiving, says the saying. Especially when a predecessor was taken out of its context (as Wright warned) and was not experienced by the evaluator but found as an image pictured in a publication. Wright Foundation Archivist Bruce Brooks Pfeiffer recalled that during the 1950s Wright often said that "Resemblances are not influences".[61]

In consideration of Barnsdall's house and the four textile concrete block houses, it was Wright's intention to make their overall external character, including the manner of relating mass, reflect the *practice* and honor the ideal of Southwest Indian buildings. They were part of Wright's experience. They were, he said, "native to the region": Wright vouchsafed they contained universal principles and architectural fundamentals and were therefore free of fashionable constraints.[62]

One aspect of Wright's romantic inclinations during this period fits the concept of a construction process for a modern adobe that might reflect the singular architecture—those "indigenous riches"—of the Indians on the high and dry, windswept plateaus and shadowed canyons of America's Southwest region. And here we find another clue as to why, at those moments around 1922, the Wrights so deliberately explored block masonry and potential aesthetic responses in sunny, quake-prone Los Angeles.

NOTES

1 Johnson (1990), pp. 97-108.

2 To add to academic confusion see *Primitive. Original matters in architecture*, Jo Edgers, Flora Samuel and Adam Sharr, editors (London: Routledge, 2006).

3 Gebhard/von Breton (1971), p. 15. On Japan and Wright see Nute (1993), but read carefully in light of this essay; and Julia Meech, *FLW and the Art of Japan. The architect's other passion* (New York: Abrams, 2001).

4 The critical relationship of Wright and his uncle Rev. Jones is an aspect of this author's on-going research.

5 FLW, *The Japanese Print. An Interpretation* (Chicago: Ralph Seymour, 1912), p. 16; 2nd ed., much altered, New York: Horizon, 1967.

6 Wright (1908), p. 161.

7 Wright (1914), pp. 411-423; Johnson (1990), pp. 32-37.

8 Johnson (1990), throughout.

9 As examples only, Alofsin (1993); Gill (1987); John O. Holzhueter, "Cudworth Beye, FLW, and the Yahara River Boathouse, 1905", *WHistory*, 72 (Spring 1989), pp. 162-198; Donald Leslie Johnson, "FLW's Architectural Projects in the Bitter Root Valley, 1909-1910", *Montana* (Summer 1987); Johnson (1990); Secrest (1992); Joseph M. Siry, "The Abraham Lincoln Center in Chicago", *JSAH*, 50(September 1991); Siry (1996); Smith (1985), pp. 296-310; Smith (1992); Sprague (1990); cf. Twombly (1979).

10 McCarter (2005), which is the second and enlarged edition of McCarter's *FLW a primer on architectural principles* (New York: Princeton Architectural Press, 1991).

11 Tselos (1953), pp. 160-169, 184.

12 The seminal essay is Tselos (1953), followed by Gabriel Weisberg, "FLW and Pre-Columbian Art—the Background for His Architecture", *Art Quarterly*, 30(April 1967), pp. 40-51. For subsequent publications with other interpretations see Alofsin (1993), p. 366,n3.

13 Tselos (1953), p. 166; and cf. Kathryn Smith, "FLW, Hollyhock House, and Olive Hill, 1914-1924", *JSAH*, 39(March 1979), pp. 15-53; Smith (1992), passim; and a balanced if unnecessarily complex essay is Neil Levine, "Landscape into Architecture. FLW's Hollyhock House and the Romance of Southern California", *AA Files* (London), 3(January 1983), pp. 22-41, abridged published again as "Hollyhock House and the Romance of Southern California", *Art in America*, 71(September 1983), p.150-156, and republished in McCarter (2005).

14 Tselos (1953), p. 166.

15 Schmidt (1978), pp. 335-351, for a good summary post-1910.

16 Alofsin (1993), chapters 7 and 8. See also De Long (1996), pp. 37-39; Levine (1996), pp. 124-127; especially Hoffmann (1992), p. 38; and cf. Tom Rickard, "Rickard on Tselos, *et al*, or Fraud in Academia?", *FLIW Update* (Newark), 6(May 1994), pp. 30-37. On contemporary written impressions of Midway Gardens cf. Kruty (1998), passim.

17 Scully (1960), after Tselos (1953), p. 25, 27.

18 Levine (1996), after Scully (1960), p. 156.

19 David Gebhard and Robert Winter, *An Architectural Guidebook to Los Angeles* (Layton, Utah: Gibbs Smith, 2003), p. 45.

20 Scully (1969), pp. 156-160.

21 Gill (1987), p. 252, see also Secrest (1992), pp. 265-270; Twombly (1979), pp. 195-198. An example of popularization within the architectural press is S.K. Lothrop, "The Architecture of the Ancient Mayas", *ARecord*, 57(June 1925), pp. 491-509.

On Southwest American and meso-American architecture as related to this essay see Bainbridge Bunting, *Early architecture in New Mexico* (Albuquerque: University of New Mexico Press, 1976); Braun(1993); Doris Heyden and Paul Gendrop, Judith Stanton trans, *Pre-Columbian architecture of meso-America* (New York: Abrams, 1973); Morgan(1881); William N. Morgan, *Ancient architecture of the Southwest* (Austin: University of Texas Press, 1994); Vincent Scully, *Pueblo/Mountain, Village, Dance* (London: Thames & Hudson, 1972); Henri Stierlin, *Living Architecture: Mayan* (New York/Fribourg: Grosset & Dunlap, 1964), esp. the drawings; Muriel Porter Weaver, *The Aztecs, Maya, and Their Predecessors: Archaeology of meso-America* (New York: Seminar Press, 1972); Gordon R. Willey, ed. *Archaeology of Southern Mesoamerica. Part One*, 2 vol, *Handbook of middle American Indians* (Austin: University of Texas Press, 1965).

22 Oles (1993), p. 165. For an exhibition of his work that traveled to Los Angeles in 1954, Wright designed a temporary pavilion on the Olive Hill campus that copied the rear wings of Barnsdall's Hollyhock house, see Aline B. Saarinen, in Meehan (1991), pp. 178-182.

23 "History galore", *Japan Times* (Tokyo), 26 November 2004, a reference to the hotel lobby rebuilt in Inuyama.

24 Hertz (1993), after Tselos (1953) and Scully (1960), p. 152.

25 Neil Levine, "FLW's own houses and his changing concept of representation", in Bolon (1988), p. 42. In another essay Hollyhock was described as a "blatant use of Pre-Columbian forms", Levine (1996), p. 140, cf. p. 160.

26 On the three preliminary elevational drawings of the Beverly Hills house see Smith (1992), pp. 40-41, 42, 167. On the Johnson "compound" see Pfeiffer (1986), plates 76-78; De Long (1996), pp. 71-79.

27 Wright (1932), p. 226; drawing Pfeiffer (1986), p. 31, where it is incorrectly dated 1913.

28 For illustrations see James (1988), plate 7, 14, 24, 27.

29 See Sullivan's National Farmers' Bank (1908), and the tombs of Carrie Getty (1890) and Charlotte Wainwright (1892) on which Wright worked.

However, why did Wright use a Barnsdall-type of architectural massing, exterior form, and architectural sculpture in Japan? And on one building, the Yamamura house of 1918-1924, why does one chimney rise on the exterior in a manner suggestive in *appearance* to combs of Mayan religious architecture? See Tanigawa (1980), pp. 41-53, 65; and "Restored Work of FLW", *Japan Architect*, no. 394 (February 1990), pp. 8-12. Construction of Yamamura began after Wright returned to the U.S. (in 1922) to construction drawings prepared by Tokyo architects Arata Endo and Makoto Minami.

30 Vernon (1998), p. 8, illus. pp. 9-10.

31 Vernon (1998), p. 7.

32 As quoted in Michael E. Welsh, "Symbol and Reality: The Cultural Challenge of Regional Architecture at the University of New Mexico, 1889-1939", in Markovich (1990), pp. 219-220

33 Maldre/Kruty (1996), p. 31, 36,n.45; Vernon (1998), pp. 10-11; H. Allen Brooks, *The Prairie School. FLW and his Midwestern Contemporaries* (Toronto: University of Toronto Press, 1972), p. 320,n.14. The New Mexico drawings are dated August 1915, i.e. before Byrne's improprieties. See also Van Dorn Hooker with Milissa Howard and Vincent Barrett Price, *Only in New Mexico: an architectural history of the University of New Mexico. The first century 1889-1989* (Albuquerque: University of New Mexico Press, 2000), pp.31-34. The university began formally in 1912.

34 On restoration <www.twboucher.com/Portfolio2/indexhtml>; see also Anne M. Nequette and R. Brooks Jeffery, *A Guide to Tucson Architecture* (Tucson: University of Arizona Press, 2002), pp. 61-245; Mark Gelernter, *A History of American Architecture* (Hanover/Manchester: University of Manchester Press, 1999), p. 200.

35 "A Hotel in Pueblo Architecture", *ARecord*, (June 1924), pp. 55-59; Onderdonk (1928), pp. 29-32; Markovich (1990), throughout.

36 One drawing in particular, 1705.031 in Wright Archive, shows straight (not inclined) attics; the date of 1917 incorrect.

37 Johnson (1987), passim.

38 Scully (1969), p. 156; see also Scully (1960), pp. 24-25; idem, "Introduction" to Bolon (1984), pp. xix-xxi, with a list wide-ranging of borrowings. On originality and/or transformation (borrowing) and/or hagiography see Diane Ghirardo and Nicholas Davis, letters, *Journal of Architectural Education*, 43(Winter 1990), pp. 62-64, idem, 43(Spring 1990), p. 64.

On the German warehouse see Margaret Helen Scott, *FLW's Warehouse in Richland Center Wisconsin* (Richland: Richland County, 1984), with thanks to the late Miss Scott; Storrer (1993), p. 185.

39 Braun (1993), p. 149. The book includes comments on buildings by Lloyd Wright that should be read in the light of Alofsin (1993) and this essay.

40 Pfeiffer (14-24), p. 88, plate 144; the structure is not a Turner system of reinforcing but the systems are similar. On Barnsdall and German buildings see Alofsin (1993), pp. 232-233, 241, 290 and related endnotes.

41 See photographs of the ornament in Thomson (1999), p. 124.

42 Tselos (1953), p. 163.

43 Bruce Books Pfeiffer, *FLW designs: The sketches, plans and drawings* (New York: Rizzoli, 2011), pp. 54-55.

44 In Wright's absence (although still providing advice) construction of the house was supervised by Russell Barr Williamson, a former Wright employee, Storrer (1993), p. 199. The drawing is illustrated in color in Riley (1994), plates pp. 148-150, and the dust jacket of Alofsin (1993).

45 The relationship of Garland and Wright is explored in Johnson, forthcoming

46 Tanigawa (1980), p. 11; James (1988), pp. 47, 22. Both books have illustrations that show the deep pocking and dramatic deterioration of stonework on the exterior.

47 Siry (1996), p. 212.

48 Viollet-le-Duc (1876), passim; *The habitations* was originallly published in Paris, 1875. When referring to meso-America Viollet-le-Duc most probably had read Stephens (1843) or later printings.

49 Alofsin (1993), pp. 221-222.

50 Johnson (2004), throughout.

51 FLW to Robert J. Goldwater, editor of *Magazine of Art*, n.d. but 1953, as quoted in Sweeney (1994), p. 234.

52 Haskell (1928), p. 12, an article Wright found to be "right", see Robert Benson, "Douglas Haskell and the Modern Movement in American Architecture", *Journal of Architectural Education* (Washington, D.C.), 37(1983), p. 7. The principal partners were architects Charles F. McKim, William R. Mead and Stanford White. Haskell was one of the few commentators that Wright—and Lewis Mumford—respected.

53 Brownell/FLW (1938), passim; Pfeiffer (2008), p. 282, where Wright's contributions are reprinted without illustrations, this particular essay was Wright's short-course history of world architecture.

54 Ada Louise Huxtable, *FLW* (New York/London: Lipper/Viking/Penguin, 2004), p. 155-164.

55 John M. Jacobs, Jr., review of FLW, *Drawings for a Living Architecture* (New York: Horizon, 1959), in *Art Bulletin* (New York), 2(June 1960), p. 166.

56 Haskell (1928), note on p. 26.

57 Wright (1931), p.49, a publication of his Scammon lectures given at the Art Institute of Chicago in 1930. They continued his ongoing denunciation of "the 'modern movement' so-called" as imported therefore unwanted: cf. Johnson (1990). Similar words were used repetitively in Brownell/Wright (1937), here and there in chapter 2.

58 As quoted by Aline B. Saarinen in Meehan (1991), p. 182.

59 As quoted in Pfeiffer (14-24), p. 88.

60 Henry Blackman Sell, "Interpretation, Not Immitation", *The International Studio*, (May 1915), pp. 79-83; a Chicago Portland Cement Company advertizement illustrated a cast concrete sculpture at Midway Gardens, p. 23

61 Pfeiffer (14-24), p. 88.

62 Wright (1932), p. 274.

7

Irving Gill, Regionalism and Concrete Adobe

The Arts and Crafts tradition and aspects of Academic Eclecticism [of the Ecole des Beaux Arts] influenced Frank Lloyd Wright, who became America's most influential spokesperson for the regionalist idea.
— Mark Gelernter and Virginia Dubrucq

Any deviation from simplicity results in a loss of dignity.
— Irving Gill, 1916

Milton Morrill was interested in the new technology of concrete construction. Within the satisfaction of his curiosity was the realization that concrete was not suited to emulating historical styles, to that form of ornamental precedent. He also realized that to effectively use the plastic yet solid material he had to simplify his buildings in floor plan and third dimension. As a result, the idea that a building should characterize a geographic social region was, probably, not even considered. Boldly he resolved that a practical approach and reductive aesthetic was necessary and sufficient. Irving Gill's motivation was different.

Gill's call for simplicity succinctly epigraphed above follows on from the words of Walt Whitman in his introduction to the 1855 and later editions of *Leaves of Grass*: "Most works are most beautiful without ornament". He was borrowing from the words of his friend Horatio Greenough who before 1850 spoke of "embellishment as false beauty". Greenough advanced that proposition by saying "Beauty is the promise of Function". In 1947 Erle Loran correctly observed that "beauty is a word sparingly used by modern creative men, but Greenough has a definition that is very easy to take". In reaction to aestheticism, the British variety, Greenough became more precise: in designing architecture, he said, "instead of forcing the functions of every sort of building into one general form, without reference to the inner distribution, let us begin from the heart as a nucleus, and [he stressed] *work outward*". He added that an "unflinching adaptation of a building to its position and use gives, as a sure product of that adaptation, character and expression". That was reduced by other people including Louis Sullivan to the aphorism "form follows function". In word and deed and before 1904 Wright's uncle, the Reverend Jenkin Jones, expressed Whitman's meaning as it was founded in Emersonian thought. Wright would refer to "organic architecture", a term woefully misunderstood since first uttered.[1] For Whitman and Greenough the greater region for creative works, for design and art, was America with an independent nationalism as ally. While Greenough spoke of "position", it is not clear if it was a building site or a locality. Regionalism, however, was another matter and credence came only with increased knowledge of North America's diverse physical attributes and indigenous cultural history and their organic artefacts.

The American Southwest "became the most vivid exemplar", the late David Gebhard has told us . . .

> of America's own ancient folk tradition. For this region, stretching from western Texas to California, possessed a number of additional elements not found elsewhere: the semiarid desert and mountain aspects of the place, the existence of settled native American folk within their own "pueblos," and the existence of a Hispanic [and Catholic] overlay. Once . . . the area had been made accessible . . . by the railroads—the South west became an internal mecca for experiencing the primitive (they thought) and the rustic.[2]

There were stimulating contrasts between the landscape and sky color of the upper Mississippi water shed and the spare dry Southwest that thrilled emigrant architect Irving Gill. The experiences of a new home-life and professional practice in San Diego compelled him to re-examine previously comforting thoughts not just about his own building designs but wholistically. Would this new almost primal environment induce ideas to fit the immediacy of present and, not by chance, reveal fundamental truths? "If we omit everything useless", he began one musing whose essence Greenough would have applauded:

> from the structural point of view we will come to see the great beauty of straight lines, to see the charm that lies in perspective, the force in light and shade, the power in balanced masses, the fascination of color that plays upon a smooth wall left free to report the passing of a cloud or nearness of a flower, the furious rush of storms and the burning stillness of summer suns.

When he spoke of the high plateaus visited he was less poetic but equally fervent when recalling

> the adobe houses of the Arizona Indians formed of the earth into structures so like the surrounding ledges and buttes in shape that they can scarcely be told from them, . . . harmonious building, . . . characteristic of their locality.[3]

Gebhard with historical knowledge and Gill by experience, express the essence of regionalism: people acting out their existence using a geographic place. Until the nineteenth century people could sustain themselves only by fashioning a life out of what nature offered: local plants and animals for food and clothing and shelter, local water and earth and stone for food and shelter. Louis XIV's Versailles Palace is an extremely sophisticated example, a hunting lodge disguised and enlarged. Yet it too relied on nature's gifts. There was no other choice. Societal traditions were constructed over time in the caldron of human history as acted out in a geographic region. Only when nature's materials could be synthesized to make new materials, and intra-region commerce flourished, were people no longer physically dependent on one place. Glass the more ancient, but iron, steel and concrete came out of nineteenth century technology and were just the beginning, not for buildings alone but the whole of synthetics. But note: synthetics are still made by combinations or manipulations of natural materials. Again, there is no other choice.

For buildings and other structures steel followed by reinforced concrete eliminated the need to construct buildings with local materials in a more than less natural state. Their natural constituents were found everywhere on the globe and then transformed by a manufacturing process that reconstituted them as new more useful and structurally adaptable materials. As such they were also non-regional, therefore potentially universal. Irving Gill understood the gravity of the situation and responded as best he could by a reductive process to simplicity. Wright understood but could not free his mind of nineteenth-century transcendentalistic and aesthetic persuasions where ornament was preeminent. His architecture, regardless of when designed, relied on traditional aesthetic responses and they included formal composition and a need for ornamentation. Only in the late 1930s did they become less essential.

IRVING GILL AND LLOYD WRIGHT

William S. Hebbard concluded a decade long partnership with Gill in 1907. Most of their many designs, to which Gill played second fiddle, had followed Arts and Crafts, bungalow, and historical styles eclectically arranged. That very year an independent Gill began to explore the meaning of simplicity as a reduction to fundamentals.[4] Uninhibited by prejudices inherent in a formal architectural education he easily included a study of Southwest Indian architecture. This was augmented by interaction with architect Frank E. Mead, an apprentice with Hebbard and Gill 1903-1907 and later Gill's partner during the single year 1907.[5] Mead had previously traveled in the North Africa of Morocco and Algeria and then west to east along the Mediterranean coast during 1900 and into 1903. He was refreshed by the experience of indigenous buildings so similar in structure and form to the Pueblo adobe with which he was familiar.

Perhaps Gill's first design to reflect in a rather mature manner his newly formed aesthetic was in 1907-1908 for the Harry Gregg houses in San Diego. While Hebbard and Gill are listed as the architect it is obviously solely Gill's design. The exterior was white stucco on masonry with no ornament and only a hint of regionalism in a trellis over a roof garden and the entry. The three Katherine Teat's and Alice Lee's "cottages" of 1905 in San Diego and the Rev. and Alice Cossitt house of 1906 in Coronade clearly were also by Gill alone and in a severe Wright Style blended with what soon became Gillian elements.[6]

As had Mead, Gill acknowledged the recent past of colonial Southern California.

> *The Missions are a part of its history that should be preserved [Gill said,] and in their long, low lines, graceful arcades, tile roofs, bell towers, arched doorways and walled gardens we find a most expressive medium of retaining tradition, history and romance.*[7]

Wright was similarly moved to observe of the Arizona, New Mexico and Southern California landscapes, that foregrounds visually spread . . .

> *to distances so vast—human scale is utterly lost as all features recede, turn blue, recede and become bluer still to merge their blue mountain shapes, snowcapped, with the azure of the skies.*
>
> *The one harmonious note man has introduced into these vast perspectives, aside from the long, low plastered wall, is the eucalyptus tree. Tall, tattered ladies, these trees stand with careless feminine grace in the charming abandon appropriate to perpetual sunshine, adding beauty to the olive-green and ivory-white of the exotic symphony in silvered gold and rose-purple.*[8]

There were, therefore, non-cultural aspects of colonial Spanish places that united buildings and landscape. In a manner that suggests an observation of Gill's architecture Wright described the buildings he found in Southern California:

> *The plain, white-plastered walls . . . of the little pictorial covens gleam or stare through the foliage in oases kept green by great mountain reservoirs. A new glint of freshness, a plainness as refreshing foil for exotic foliage.*

And there were "cool patios".[9] And "Buildings could grow right up out of the ground as adobe buildings do. . . ."[10] This was a place completely opposite to the blue-green rolled forested and humid-laden vales and frosted white winter plains of the northern midwest; even more opposite to the cluttered cramped damp green Yucatan jungles.

The observations of Wright and Gill fit two universal human experiences: that a person's senses are challenged, invigorated and then refreshed by new, exotic environments, and, that non-indigenous migratory influences are adopted by people of a region resulting in hybridizations.[11] Gill's Southern California architecture is a wonderful case of hybrid vigor. We can say the same of Wright's four textile block houses but not the theatrically insipid Hollyhock. But note: Wright's observations are akin to Lloyd's. In turn they recall quite closely those made by Gill much earlier in magazines the Wrights regularly read and, as a certainty, in friendly conversations. As a practical example, historian Thomas Hines observed that, "Except for his characteristic green trim and his dark reddish-brown concrete floors, waxed to the texture of 'old leather,' Gill's mature buildings were painted exclusively in warm shades of white, inside and out".[12] White: a neutral universal generic color. Many of the concrete floors of Wright's houses after 1930 and into the 1950s were painted a color he called Cherokee red that was highly waxed. Where used, textile block walls were left the grayish dry white of Portland cement.

Artists Bertha H. Smith and Eloise Roorbach, two of Gill's San Diego contemporaries and "most insightful early interpreters", wrote short laudatory pieces. Smith was certain that Gill "the architect",

> [took] delight in color and would teach you to find it as he does, in the reflected glow from the red floor of an open court, bank of flowers, a green terrace in shadows cast by a curtain of vines, in all the varying lights of day and evening as they call from those walls the infinite hidden tones of the painter's blending.[13]

To Roorbach's theoretical delight Gill "artfully embodied the permanent principles in the straight line and circle".[14] To her aesthetic delight,

> The rooms seem tinted with delicate pastel shades one hour, become iridescent at another, are dove gray overlaid with rose at another. To live in such a room is almost like living on the inside of an opal or in the heart of a flower.[15]

Historian's Hines's own experiences of Gill's buildings were expressed in equally romantic language. Unlike Wright or the Green Brothers,

> or various figures among his European contemporaries [Hines said], Gill was decidedly not committed, except in several of his Arts and Crafts houses, to the idea of the . . . totally regulated and centrally coordinated work of art. Instead, he designed benignly plain, white, neutral, functional, light-filled buildings into which users were invited to bring their own personal lives. . . . Spare, selected pieces of antique furniture were welcomed, as were more contemporary pieces by such Arts and Crafts designers as Gustav Stickley and his brothers.[16]

Gill's houses were logical in plan and in elevation plain but not spare, elemental and simple and homogenous, see Fig. 7.1. Like Gill, Mead, and son Lloyd, Wright now wished to create a modern architecture using as inspiration, and that seems the correct word, as inspiration the utter clarity of *structure* and *form* as found in Southwestern adobe and rock buildings and in Gills' modernism.

In spite of their apparent visual differences we shall learn that Gill and Wright were not regionalists. Rather, from the region they cultivated insights that would lead to theoretical expressions. For Gill that cultivation included assimilating Mead's experiences in Northwest Africa, a geographic environment similar to places inhabited by American Southwest Indians. Clearly, though, Gill was a rationalist, Wright an aestheticist. But Gill's expressions were full of potential, not encumbered by idiosyncrasies as the Bella Vista Terrace housing at Sierra Madre in the San Gabriel Valley clearly demonstrated in 1910.

> These buildings of California, [Roorbach said], consist of separate houses placed side by side as flowers are placed in a garden. . . . The whole [city block] plan is strikingly original as to treatment of a given space, style of architecture, and construction. . . . [S]implicity carried to the last word in architecture art.[17]

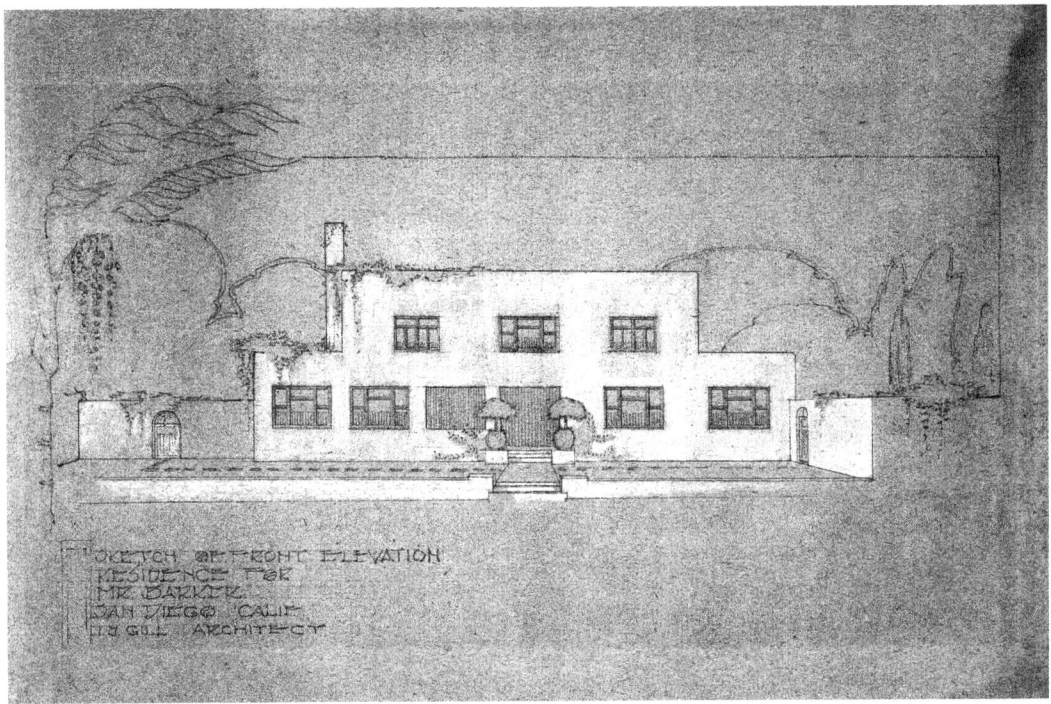

Fig. 7.1 Nelson E. Barker house in San Diego, California, Irving Gill, Architect, 1911-1912

Wright's cloyingly sentimental Barnsdall house was an unsatisfying first attempt at responding architecturally to the region and its pre-Columbian history.

The measure of father's influence on Lloyd will always be problematic; so too Lloyd's upon father. That of son John Kenneth—who preferred John Lloyd—Wright, was less apparent and is difficult to pin point. This can be inferred in his writings, even those about Father as his god on earth. When father left for Europe in 1909 John was completing the final year at his great-aunts Hillside Home School. He attended the University of Wisconsin in Madison for one year, leaving disenchanted. During his 'teen years John attended design studios at the Art Institute of Chicago, was a draftsman in San Diego for a building company in 1912 and that year began with the successful commercial architect Harrison Albright, an advocate of reinforced concrete. During a brief tenure with Albright John worked on plans for the Golden West Hotel, a modernist design in concrete and in contrast to Albright's historicism, and brought in the sculptor Alphonso Iannelli who later worked with Wright Senior. John then worked for father late 1913 into 1918 in Chicago—where he concentrated on working drawings and superintend construction of the concrete structure Midway Gardens—and in Tokyo. After 1918 he established a practice in northern Indiana and Illinois, and only occasionally assisted his father.[18]

Lloyd studied engineering and agronomy from 1907 to 1909 at the University of Wisconsin in Madison. He began working in 1911 for landscape architects Olmsted and Olmsted, first in Boston and then in San Diego where he worked on the Panama-Pacific Exposition at Balboa Park. The Olmsted firm had been brought to San Diego on the advice of Irving Gill, a family friend, but in 1912 Olmsted resigned from the exposition project. The Wright brothers lived briefly in an Irving Gill-owned and designed cottage.[19] Lloyd began in Gill's office in 1912 and shortly thereafter they moved north to open a second office in Los Angeles. Apparently Lloyd remained with Gill until 1914.

The professional paths of Gill and Wright Senior mingled, socially they crossed. Wright worked in the Chicago office of Joseph Silsbee 1886-87; Gill during 1890-1891. Gill and Wright were together in the Chicago office of Dankmar Adler and Louis Sullivan from 1891. Both were laid off in 1893 because of the sudden and crippling economic recession of 1893. Wright then set up offices in downtown Chicago and in his Oak Park house; Gill moved to San Diego.

Beginning in 1916 Lloyd worked as a movie set designer at Paramount Studios to then return to places in and around New York City in 1917. Probably in that year he married the Chicago actress, muralist and printmaker Kira Markham (nèe Elaine Hyman) who had been professionally associated with Aline Barnsdall in New York and earlier with Chicago theater and Theodore Dreiser. Manhattan not to their liking, smitten with things Hollywood and looking to work in the motion picture industry, they settled in Los Angeles in November 1919. After 1923 Lloyd, now divorced, was occupied independently as a part time free-lance landscape designer while involved with the odd movie set design (as had Kira), mostly for Paramount studio. It was not until 1928 that he received a California architect's license.

In 1913 Wright visited sons in San Diego and Los Angeles and naturally talked with Gill. In 1915 he again visited his sons and Gill while also attending the San Diego Panama-Pacific Exposition that celebrated the opening of the Panama Canal. San Diego wanted to emphasize that it, not Long Beach or San Francisco, was the first and principal U.S. west coast port for ships moving east to west through the canal and north. Architecture for the Exposition came under the control of architect Bertram Goodhue whose stylistic preference was for a restrained and well-executed Spanish baroque; some people might refer to it as a revival. Gill was initially considered to direct planning and architectural activities, but the more socially attractive and architecturally fashionable Goodhue was placed in charge. For reasons yet known he then excluded Gill from participation.[20]

John Wright has stated that southwestern "Native" architecture had the "solid integrity of America" and that Gill "tuned in to" that same "vital Source" as had John's father.[21] We've noted the truth is vice versa; Wright "tuned in" well after Gill's mature architectural designs of twelve (12) years earlier. To John's reckoning Gill was an "unsung hero", "moonlike", a "courageous creator" who began developing a new, refreshing architectural style about 1904, to be maturely clarified by 1907.[22] "Gill looked after me like a son", Lloyd has said, adding that Gill "wasn't a bookkeeper; he was a dreamer and poet;"[23] he was a man of "great charm", see Fig. 7.2.

While is southern California 1913 and 1915 Wright experienced Gill's buildings. When together men's-talk must have centered on architecture. Wright recognized that Gill did "carry on a line of experimentation along his [Wright's] own lines", and had "developed a technique more 'adobe' and Indian" than his own.[24] In 1932 Lloyd clumsily said, and father agreed, that Gill . . .

> *[was] The most original of all the architects living on this coast. . . . [He] worked alone here [in Southern California] . . . to create simplified structure*

Fig. 7.2 A "cube house" project by Irving J. Gill, Architect, of ca.1910

in a land of heat and dust[,] of suitable materials, plaster, concrete, and cement. Sanitary and Cool. . . . Alive to his environment and what it meant[,] and to the changing times[,] in his efforts toward standardizing and simplifying structure and assemblage.[25]

Lloyd was referring to Gill's use (now and then) of the cost saving concrete tilt-up slab construction that he first employed on designs for the Banning house in Los Angeles and then the La Jolla Woman's Club in 1913.[26] Among many others historian Mitchell Schwarzer not incorrectly observed that,

The recently discovered technology of reinforced-concrete construction was the [or one?] basis for Gill's new aesthetic. . . . Using either a concrete frame with hollow terra-cotta tile in-fill or poured-in-place concrete walls [i.e. tilt-up], Gill began to experiment with very thin walls, bringing down a building's mass to its utilitarian essentials and bringing down its cost to middle-class affordability.[27]

But most of Gills' buildings were constructed of hollow clay masonry walls that were plastered and stuccoed, concrete floors and timber roofs, .

Schindler came to know Gill during a 1915 tour of the southwest that included San Diego. Schindler then used tilt-up concrete panels on his Kings Road duplex house in West Hollywood in 1922. Therefore Lloyd was also referring to aesthetics, to Gill's precisely reduced amalgam of a severe planer abstraction of the utmost clarity. In no manner was Gill's post-1906 architecture dependent on historical eclecticism except as he might introduce the occasional round arch. And it was not technology driven. Inspired by Sullivan's "luminous idea of simplicity", as Gill put it, he believed that:

There is something very restful and satisfying to my mind in the simple cube house with creamy walls, sheer and plain, rising boldly into the sky, unrelieved by cornices or overhang of roof, unornamented save for the vines that soften a line.[28]

That is not historicism or regionalism. Architectural historian Thomas Hines has put it well.

> Like most architectural rationalists, he [Gill] eschewed historicism and strove, in his mature period, for pure, new, and original statements, effecting a "modern life".[29]

Wright Senior acknowledged this when he said Gill's designs were not infected by the plague of sentimentality, rather they had "antiseptic simplicity".[30] (So had Morrill's.) Gill's all-concrete Walter L. Dodge house on Kings Road in West Hollywood (1914-1916, demolished 1970), where building masses are juxtaposed in a manner reminiscent of Pueblo housing, was only a few doors from an innocuous house Wright rented from 1919 to 1921, and near Schindler's own Kings Road house begun in 1921.

Gill's was a clean, uncomplicated, white-ish architecture that predated Adolf Loos's houses in Vienna by three years; the French architect Tony Garnier's housing as revised and published in 1918 (with some housing designs similar to Morrill's and Gill's) by twelve years; and by nearly two decades the mechanistic buildings the Dutch and German modernists settled upon in the early 1920s. Wright was fairly certain of this when in the 1940s he recalled those early days in which he played an important role, saying that,

> Gill was a rank an individualist as I. . . . But his individual character came out to good purpose in the good work he did later [after Chicago] in San Diego and Los Angeles. His work was a kind of elimination which if coupled with a finer sense of proportion would have been—I think it was, anyway—a real contribution to our so-called modern movement.[31]

Lloyd believed that "Gill felt that in his structure[s] he had the support of men in Europe who were also going toward simplification".[32] It was a tentative comment probably made in the 1920s and may have referred to Loos or Garnier or even Le Corbusier.

Worried about Austria's cultural well-being in an age technical advancement in a socially and politically disturbed Europe, and in search of a proper modern design response, prior to 1910 Loos was not well known beyond home country and Bavaria. Further and specifically, while a series of short lectures denouncing ancient or fancifull ornaments as cultural crimes were given in Vienna beginning in 1908, *Ornament und Verbrechen* ("Ornament and crime") was not published for wider consumption until 1912, next in 1929 and then 1926, each in Paris, then in German in 1929, revisions most likely.[33] Any thought that the essay or Loos's interior designs had an influence upon Morrill or Gill or other American architects before 1913 must be rejected.

Loos was aware of Sullivan's work (Wright then his employee) as a result of a visit to the United States in 1893 and a wander about Chicago's streets and baroque World's Columbian Exhibition. He also visited and worked not as an architectural assistant but at odd jobs in New York, Philadelphia and St. Louis. After returning to Austria in 1896 there was no immediate aesthetic or stylistic design response by Loos to America's architecture. But in time he assimilated his experiences and formulated thoughts that might stimulate Austria in the new century. As part of the artistic and intellectual elite he nonetheless was intent on introducing

American civilzation as a useful paradigmatic pragmatism for Austria. To unshackle a hidebound society through exclamation in the arts he directed two main themes: "The work of art is born without any existing need. A house fulfills a need" ("art is useless", Oscar Wilde, and architecture is a design process not art), and two, the absence of ornament is a sign of spiritual and cultural strength.[34] Was Loos aware of Greenough's writings or those theorists of similar mind. Or had he read Whitman who was very popular and a favored author of Sullivan, his followers and crafts-oriented designers and artists who were everywhere where went Loos in America.

As a result of publication in 1910 of Wright's work in Berlin by Wasmuth, a folio and a book well known in Vienna, one Wrightian element can be see in plan *only* for Loos's compact house located on Northartgasse, Vienna, of 1913; a floor plan and building so un-Viennese.[35] That aside, the critical observation to savour is that just as Gill responded to architect-partner Frank Mead's tales and photographs of North African vernacular buildings and to Gill's own experiences with New Mexico aboriginal buildings, Loos was influenced by Algerian vernacular architecture. This was gladly acknowledged when describing his design of the Scheu house in Vienna of 1911-1912.[36] This house is architecturally superior and of greater theoretical value to his oft-mentioned and so ill-proportioned Steiner house of one year earlier. In fact Loos's first built *building* was the Karma house near Lake Geneva of 1904, second the Steiner house of 1910 and next the Michaelerplatz Building of 1910-1911 followed by the Scheu house, these in Vienna.[37]

The influence or not of Morrill's and Gill's work on European architects prior to the military hiatus of 1914 has yet to be studied. The use of American design books and magazines by European architects or the route of their North American travels during the years, say, 1893 into 1913 are to be adequately discovered and then analyzed. For the present it is safe to say there was a simultaneity of stimuli, theoretical and experiential, that induced the reductive design philosophy presented in theory *and* practice first by Gill, followed by Morrill, then Loos; Gill's and Morrill's the more usefully ascetic.

There is an interesting anecdote. At one moment Barnsdall mentioned that Wright envisioned a white house (a la Gill?) sitting brightly on Olive Hill. However, she preferred something less contrasting, more subtle, and prevailed with a sand color of "light gray-green", much like the natural, blended earth-colored pueblos.[38]

BARNSDALL'S PUEBLO

If a historian is uncertain about visual or experiential evidence, there is the verbal, so dear to those who need written confirmation of obvious stuff. And Wright satisfied those uncertain people, if they had but read and looked with an open, inquiring mind. Soon after Hollyhock was completed, in his typically awkward prose Wright wrote in an article that . . .

> *The Olive Hill work . . . is a new type in California, a land of romance—a land that, as yet has no characteristic building material and no type of building except one carried there by Spanish missionaries in early days. . . . I feel in the silhouette [we stress] of the Olive Hill house a sense of the romance of*

Fig. 7.3 House project for G.P. Lowes at Eagle Rock, California, 1922

the region when sens[e] associated with its background and in the type as a whole a thing adaptable to California conditions.[39]

He was right about Southern California. Being food gatherers in such a mild climate, Indians, Chumash mostly, had no need for permanent buildings. Grass shelters as wind breaks and to slope off rain were sufficient. Wright illustrated the article with pictures of Hollyhock and the Lowes house project. Drawings by Schindler and Lloyd dated 1922 reveal the Lowes house, the second design so intimately similar to Schindler's "monolithic" design, was conceived as a series of plain stuccoed cubic forms rising sun bleached out of the earth, see Fig. 7.3. Wright lost the commission, but after a few months and with slight modifications to floor plans, the sand-surfaced cubic forms were transformed in 1923 to become the textured Storer house.[40] The client's program and site therefore did not lead to the architectural concept, let alone the floor plans or to the use of textile blocks.

A couple of years later Wright said that Barnsdall's Olive Hill buildings embodied the "characteristic features" of the American Southwest region and opined that Hollyhock was a "natural house, naturally built; native to the region of California as the house[s] [he had designed] in the Middle West had been Native to the Middle West".[41] In other words he now believed that his buildings were a regionalist's response. Before that concession it was perfectly all right to design a Wright style house typical of those for the northern midwest prairies with exterior walls of stained board and batten for the Stewarts in 1909 at Montecito, nearby to Santa Barbara in California. It was in all respects similar to George and Alice Millard's first Wright-designed house of 1906 for site in the Highland Park suburb of Chicago, see Fig. 7.4.

Fig. 7.4 The George and Alice Millard house in Highland Park, Illinois, 1906

In the 1970s son Lloyd reconfirmed that, in summarizing his own words uttered in 1931, "his father's conception of the [Barnsdall] house . . .reflects the Southwest", that some people had "called" Hollyhock

> modern and Mayan; but no one realized what father was giving them. He had submerged himself into the area in spirit and developed a true expression or architectural characteristic of the Southwest—his Romanza. What he had built was a mesa silhouette, terrace on terrace, characterized and developed by Pueblo Indians, [the house] "reflects the Southwest".[42]

Interestingly, "Actor's Adobe" was a title the Wrights gave in 1919 to preliminary plans of living quarters for people who would be involved with Barnsdall's proposed artists' colony on Olive Hill. A U-shaped floor plan had a covered ambulatory on three sides of an open court, a scheme so very typical of rancho or hacienda or other moderate Spanish colonial buildings in the region. The walls were to be bricks of sun dried mud and timber.[43] But like so much of Wright's output during this period it remained sketches.

The meso-American pre-Columbian fiction was (until now) sustained by a pursuit of what historian Alofsin quite correctly referred to as "superficial visual analogies".[44] That hunt was one lingering aspect of late nineteenth-century art-historical methodology that grew awkwardly out of the new and burgeoning discipline of archaeology. In contemporary practice one aspect of the method relied on visual comparisons of buildings often pictured in publications (so seldom experienced by an author) that illustrated older buildings and their external elements and elevational composition. Superficially then, by visual comparison of the exteriors and their ornament (almost exclusively), possible progenitors are identified. Just as archaeological research methods have matured dramatically during the last ninety years, so has architectural history.

The experiential evidence is unequivocal for the observant viewer, student (scholar) and user. Further, when in the early 1920s Wright selected and then sent photographs of the Barnsdall house to Hendrik Wijdeveld in Holland for publication in the Amsterdam magazine *Wendingen*, a number of them conveyed the "characteristic features" of collected boxes and roof terraces at Taos or of Zuni mesa villages, of buildings in repose beside or on a hill and made of the hill.[45] Those

Fig. 7.5 Barnsdall house in Hollywood, California, 1919-1921

photos, amateurish and perhaps by Wright, pictorially portrayed exactly Lloyd's verbal statement and his father's admissions.

Also, the canted or battered attic wall, albeit not as tall, can be found as a parapet (or railing or balustrade) on some adobe buildings by Native Americans, and on Spanish colonial—and their imitative—buildings.[46] A parapet usually began above *vigas* where non-structural material was required. Virtually the only difference in external *appearance* between the ancient desert buildings and Hollyhock, set in verdant grasses and bushes mixed with acres of introduced olive trees, was the uneven wall surface of the old, and the sharp corners of the new. Hollyhock's walls were stucco uniformly troweled onto hollow clay tiles (perhaps Natco's) and onto wire mesh tacked to timber frames and onto commercially available concrete blocks.

Hollyhock's general plan configuration also recalls, if more formally, the Native American adobe buildings of the Southwest as they haphazardly gathered about a ceremonial space, where roofs were used for participation in or observation of ceremony.[47] There are a number of roofs at Hollyhock (with parapet), that focus on the grassed courtyard *and* on a raised central stage. The stage's rear wall, clay floor tiles, and an absence of a safety parapet facing the courtyard define the stage area. As at Pueblos all roofs are accessed by exterior stairs. In Fig. 7.5 stairs from the grassed courtyard (in the foreground) lead up to the stage located on the roof of the loggia. Also, there are steps that run the full width of the garden courtyard that lead from the grassed yard down to the loggia floor level. The courtyard, therefore, is raised a few feet so a standing or seated audience could better see activities on the rooftop stage. Apparently a few musical performances were presented on the

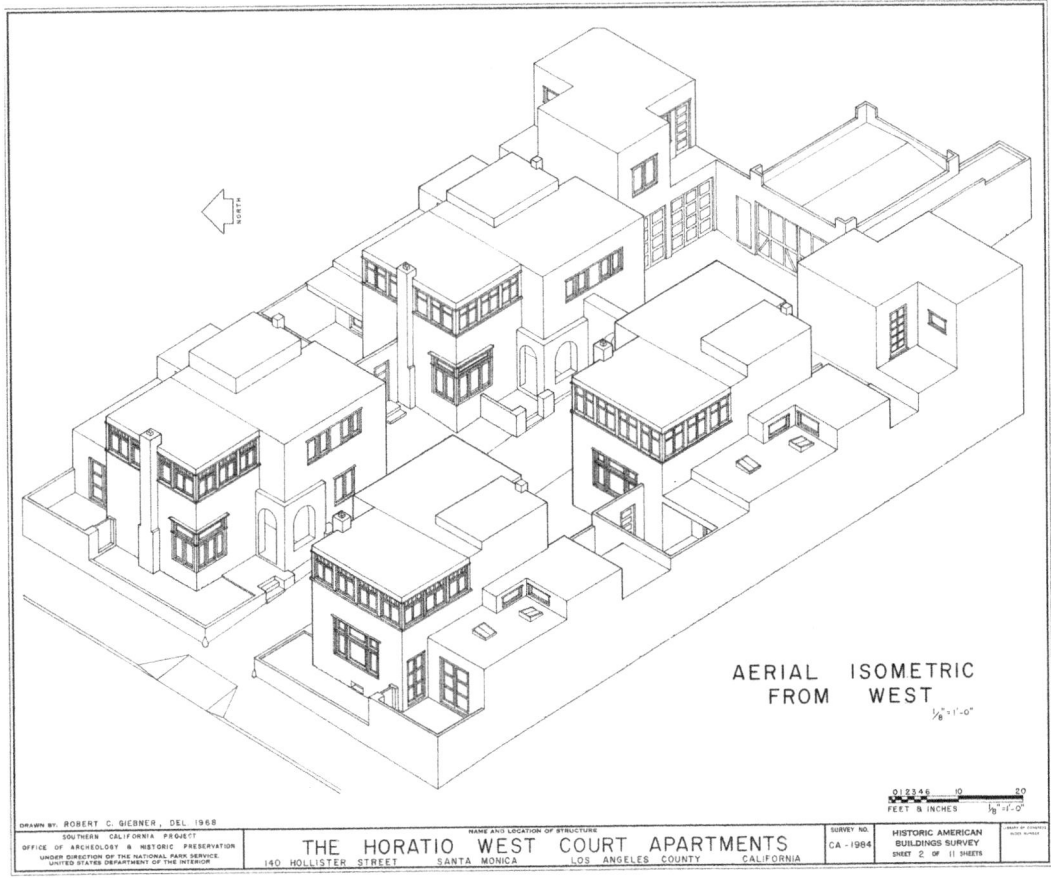

Fig. 7.6 Horatio West Court in Santa Monica, California, Irving Gill, Architect, 1919

stage but there is no record of a drama production. A couple of the other roofs were initially covered with grass turf imitating natural ledges.[48]

Theatrical producer, director and theoretician Sheldon Cheney was one of Barnsdall's advisors, a role shared with Eleanor Duse who was Barnsdall's philosophic mentor. In 1918 Cheney said that "drama was closest to the life of the people [when] it has had its setting in the open", in "the out-of-doors". He also quoted Duse who, with the example of ancient Greece and later wandering players in Europe, believed that drama should "play in the open air".[49] The venerable Dutch architect H.P. Berlage thought Barnsdall's Hollyhock was influenced by Japanese art (apparently he saw oriental motifs) and referred to its "open-air theater", a bit of information he must have got from Wright.[50] Miss Barnsdall and Wright *intended* the grass-and-tiled roofs and grassed courtyard be used for soirées and theater; like the Pueblo people.

Wright took more than inspiration from the pre-Spanish architecture that had evolved within the high desert region. Indeed, with exceptions in the 1890s, during the span of a long professional career, Hollyhock was as close as he would come to imitation. But that was *only* the exterior massing where the finish materials and color enliven the innovative emulation. The interior was almost totally derived from designs made in the good old days of the Wright School of the Midwest.

Just as Schindler and Lloyd and Wright were finalizing plans for Barnsdall's Olive Hill mansion Gill's revolutionary and prescient Horatio West Court Apartments in Santa Monica were nearly finished, see Fig. 7.6 opposite.

INDIGENOUS

> *Regionalism in architecture is the desire to shape buildings according to the particular characteristics of a specific place.... [It] aims to obtain a visual harmony between the building and its geographical setting, to connect with the culture and architectural traditions of the local area, and to manage the local climate as naturally and as efficiently as possible.*[51]

And so on. Wright was not interested in theories of regionalism. His responses were based on first and second hand experience: that was sufficient. Perhaps that was best because if he gave it more than a smidgin of philosophical thought he might have reached conclusions similar to those of a present-day practitioner in New Mexico, architect Glade Sperry, Jr.

> *It is an alluring trap for those of us who must create in the shadow of its [the region's] force.*

Although it is possible to build anew in the materials that have been used in the region for hundreds of years, it seems somehow limiting—a narrow view of the possibilities. The architecture of this place is truly a part of this place; it seems to be rooted in the land in a primal way. Forms are not manipulated; they grow in response to a sense of organization that has much to do with survival in the desert.

> *The softly focused adobe walls, hand-carved vigas, and patterned brick floors entice you with a familiarity that can deaden creative thought. They do not portend new opportunities for the spaces we inhabit. To replicate the evidence of the culture that came before dooms the culture of the present.*[52]

During the five decades before 1900 northern Euroamericans were enthralled by the traditional land owners'—the indigenous Indians'—exoticism, their life-styles and unique arts. Yet, their land was assumed free (a kind of *terra nulius*) and open to Euroamerican settlement and exploitation. Indeed, Theodore Roosevelt, in chilling prose, offered the then-persuasive view that the Indians had no legal title to their land, that it belonged to the "conquerors".[53] Acquisition of the traditional land owners' realm and domain west of the Mississippi River was purposeful, if isolated and sporadic, before the Civil War. But with rail lines moving west out of Illinois, Kansas and Texas to serve homesteaders, cattlemen, miners, and land speculators, a frenzied land rush was on. In a strange twist in history, by the 1880s, a period during which their cultures were almost torn asunder, all things Indian became even more popular with Euroamericans, to remain so.

A few anecdotal examples make the point. Fred Harvey opened his first restaurant in 1876 and then aligned his company with the Santa Fe Railroad. By 1901 he had established fifteen hotels and forty seven restaurants, many in New Mexico, at the Grand Canyon, and in California. Beginning in 1902, architect Mary Colter "designed hotels, train stations, shops and restaurants for Harvey, steeping them in Native American history and art".[54] Natural building materials were the dominant feature of her buildings. In 1885 Sitting Bull joined Buffalo Bills Wild West (started in 1883) and the show played Chicago. The newly created U.S. Bureau of Ethnology sent its first expedition to the Southwest in 1879. A member of the field team was Frank H. Cushing. On arrival at Zuni Pueblo he decided to remain in residence in the belief that to understand the Zuni you must live and share with them. The Zuni found him a sincere anthropologist who became fluent in their language.

It was not to academic or learned journals that Wright would have been attracted, but to articles in popular magazines like *Atlantic Monthly* and *Harper's*. Cushing's "My Adventures in Zuni" was serialized in *Century* during 1882 and 1883. Or he read books like the popular (and highly acclaimed) *The Cliff Dwellers of Mesa Verde* (1893) by Swedish naturalist Gustaf A.N. Nordenskjold or Cushing's *Zuni Creation Myths* of 1896.[55] Or he attended the Middle America Exhibition of full-sized replicas of Puuc Maya structures at Chicago's in 1893 World's Columbian Exhibition. It celebrated a period of considerable archaeological work, one benefit of which led to the founding of many museums and research institutions. Among them was the Field Museum of Natural History in Chicago. Pueblo, Hopi, Navajo, Azetc and Maya displays at the Chicago 1893, St Louis 1904, and San Diego 1915 expositions were attended by Wright and his entourage who found them a critical and broadening encounter. A pueblo housing structure of three stories was built at St. Louis with Indians in traditional dress in attendance. However, Wright did *not* react to those stimuli in designs until after his 1915 visit with Gill!

Also to make newspaper headlines in 1890 was the massacre by the U.S. Army of 250 Lakota children, women, and men beside Wounded Knee Creek in South Dakota. Invited by the Army as witnesses, the press look on and then published reports and photographs for Euroamerican consumption. It was another retaliation by the Army for Custer's humiliation at Little Big Horn and the last major battle of the infamous Indian wars. But it was not the last display of feigned ignorance and destructive exercises by maladministration in Washington, D.C., or in state and territory capitals, notably Denver.

In California there were many tribelets living in the area from present day Monterey south. They were food collectors and built ephemeral bush, reed and grass shelter.[56] When Euroamerican settlement began, from San Francisco south there were 200,000 Indians, all of whom offered welcome. In 1890, in the wake of murdering hunting parties, famine and European diseases, no more than 29,000 were counted. In that year Chicagoans began planning their World's Colombian Exposition extravaganza for 1893; Idaho and Montana were admitted to the Union; Wright was working in the Adler and Sullivan office helping create skyscrapers; men luxuriated in city clubs; the U.S. government obtained 11,000,000 acres of legally reserved Sioux land; women worried about servants, shoes and bustles or joined civic groups; children worked in factories; the University of Chicago was established; railroads and roadways everywhere were long scars on the earth; Yosemite National Park was created; a flood of foreign immigrants pushed the U.S. population to 62 million plus; *How the Other Half Lives* by Jacob Riis was published; and on and on. And the ill-defined frontier dividing native and conqueror inexorably shifted and constricted the Indians.

Prior to 1900, therefore, there was indelible and high contrast between industrialized, mechanized and urbanized Euroamerica and primal, earthbound, yet sophisticated Indian cultures. That distinction was most obvious in the unique Southwestern community buildings set in—and made of—high plains earth and stone. It was understandable for Europeans and Americans to be curious about those exotic people.[57]

Son Lloyd Wright "quite openly referred to his buildings as art objects which sought to convey the spirit of the American Indian".[58] He was intrigued with the colors, vegetation, and adobe architecture of the Southwest as well as buildings designed by Spanish conquerors that around 1900 became known as Spanish or Mission revival. Lloyd proudly admitted some of his designs were influenced by plains wigwam or tepee forms. There was, for instance, his first Hollywood Bowl of 1924-25, and the Dune house project of 1927.[59] His finest works of the 1920s and 1930s were those inspired by ancient buildings found in the wider Southwest region. His better buildings included the Bollman house where the front elevation was so similar to the street facade of the later Barnsdall Director's House on which Lloyd and Schindler worked together for Wright Senior. The remodeling of Barnsdall's Residence B, begun in 1923 when only Lloyd was involved, on one facade only were many characteristics of the Bollman house and Lloyd's own studio/house of 1927.

Lloyd has said that certain of his own designs and verbal suggestions were taken on by his father: one was knit-block. Anyone who has worked in an architect's office will confirm the give and take, and understand the degree and direction of persuasion. The Wrights' must have discussed not only immediate things architectonic but subjects such as Indian arts, primitivism, space and landscape, Americanism, color, arts and crafts, and all else including Gill's architectural interpretations of which they were so fond.

There is a clear visual connection to northern plains Indians in Wright's preliminary drawings and proposed sculptures for the Madison Realty Company's Nakoma Country Club of 1923 and 1924. The hills of Nakoma near Madison were occupied by local Indians, perhaps the same people, the Algonquin or Winnebagos, the Joneses had displaced in the 1850s when claiming land for their farms. Indians were still living in the area when as a teenager Frank spent a couple of summers on the Jones family farms and when his aunts began their progressive school on family farm land. The reality company had used the services of prominent landscape architects Ossian Cole Simonds and Leonard Smith; both laid out streets and allotments at Nakoma. Over about twenty five years Wright's old friend Franz Aust executed a few additional landscape plans.

Wright's design for the clubhouse contained a plan of complex shapes sprouting out from a hexagonal hearth. All was covered by a complicated series of roofs, some of which were teepee shaped structures and, for the central lounge and hearth, apparently derived also from the typical grand national park buildings of the era. The walls were to be random ashlar stone, not textile blocks. In the end they chose another architect who designed a clubhouse that appeared a bit like a large West coast bungalow. Wright was also asked to design a Memorial Gateway to the Nakoma suburb. He designed sculptural images of an Indian male and female that were to stand beside pools of water either side of the road entry. The two figure designs were eventually built as sixteen foot high sculptures for placement in a courtyard of the Johnson Wax Building in Racine, Wisconsin, designed by Wright in 1936.

The connection is also clear in proposals for mountain cabins (one to have been in textile block he called the "Big Tree Wigwam)" for his own unrealized real estate venture for a Lake Tahoe summer colony in 1923.[60] They were apparently designed to employ textile blocks or tiles but exactly how is not clear from documents. During those difficult years there was the architectural commission for Mrs Gladney that in style and detail of wood siding and concrete walls (not blocks) belonged with the series of hypothetical buildings designed for Wright's ill-fated Lake Tahoe colony. But her house was not built.[61]

The Nakoma, Tahoe and Gladney designs also drew inspiration from the many log, timber and stone lodges that had been (since 1890)—or were being—constructed in national and state parks and other recreational areas such as Yosemite, Mt Rainier, and Yellowstone. Was Wright attempting to distinguish Indian characteristics by the region to which a building was proposed and then by association to modern; or at least the contemporary?

While work on the Nakoma designs was drawing to a close, the German architect Eric Mendelsohn spent a week-or-so vacation with Wright at Spring Green in October 1924. Former Wright employee Francis Barry Byrne was an invited guest. Mendelsohn recalled that one afternoon everyone "had to change" into Indian costume, "Bark shoes, a long staff, gloves and a tomahawk". They then hiked and picnicked on land "abandoned by the redskins". During the visit Mendelsohn made a sketch on which, he said, "I had to sign it and write: 'To the Master'".[62] Was it also a game when Wright and author Hamlin Garland costumed themselves as Indians two decades earlier? Can too much be read into what seems to be a frivolous activity?, almost childs-play? The answer is perhaps. But consider this: by the 1920s there were in the U.S. about 800 "fraternal orders that incorporated Indian lore and costumes in their secret ceremonies".[63] Indeed, things Indian were very popular with Euroamericans emigrants but their response was for the most part superficial, trivial, demeaning. At Nakoma there was Mohican Pass, Oneida Place, club members were "warriors", golf greens named Many Scalps, Papoose, and so forth. Oddly, "Nakoma" is a Chippewa word meaning "I do as I promise". Another Madison golf course was called the Blackhawk Country Club.

As newly arrived migrants to Southern California Wrights' clients must have anticipated a regional architectural response when they considered the sites of archaeological fragments of past (and living Indian and Spanish communities) and vegetation and vivid in the sun. It was so vitally different to the wet green northern places they had left. They were as enthralled by the area's geographic and anthropological attributes as were Mead, Gill, Schindler and the Wrights. Culturally and cosmetically different to their childhood experiences it seemed so seemingly unadulterated, sand colored under bright blue skies, clean-cut, apparently undefiled. Father and son agreed that the "standardized concrete construction" used for the "Millard residence in Pasadena" was, to quote, "designed specifically for California and inspired by the local environment".[64] Yet, the houses were, we must accept, not as Barnsdall, Millard, and the others could have anticipated. Storer move out of his house after one year. Perhaps he could not shed the psychological baggage and experiences of house from such a demanding and insisting architectural enclosure for home.

CONSOLIDATION

When a young professional Wright was not keen to visit Europe. It was a place of known architectures that he and Sullivan found undesirable for the Europeanized new world. When he took time to travel abroad around 1905 it was to Japan with a brief side-trip to China. Those were first experiences of exotic people and their art and architecture. Except incidentally, he has said nothing about the Chinese architecture he experienced, but he and others have said much about the Japanese. In 1928 he was emphatic: "I knew nothing of Japanese architecture until I first saw it in 1906", actually 1905.[65] That is to say, understanding came with personal *experience*. Two dimensional art is another matter. He candidly stated that of his presentation drawings made before 1911, notably a buildings' perspective, "Their debt to Japanese ideals, these renderings themselves sufficiently acknowledge".[66] However, he did not execute one presentation drawing: all were by employees or commissioned.

When he finally traveled to Europe in 1909 it was to accompany his lover Mrs. Mamah Cheney who took up a short term position in a German university. During those months he prepared drawings for the publication of his architectural work by the Berlin publisher Wasmuth. He also visited buildings by the modernist Joseph Olbrich and other like-minded Viennese designers and some of the decorative and sculptural designs found in northern Italy and Germany. Wright appreciated that these contemporaries were struggling to shed the impress of centuries of Greco-Roman and Gothic styles and design elements whose physical remains were found everywhere.[67] His one year study leave in Europe affirmed Sullivan's and his theoretical position of the frustration, the shear poverty of traditionalism.

The American Southwest indigenous or Native architecture was his next experience of exotic places and people. It stirred his creative imagination and eventually tempted him to relocate there at a time when his personal life was in turmoil as a result of the murder of his lover and her children, followed by an affair with an awkward mistress who was a morphine addict. This was coupled with declining professional fortunes in Wisconsin and Illinois, and possible isolation and insulation in Japan.[68] It all come to a head in 1922 when, even before the Tokyo hotel was finished and with no new commissions in Japan or the United States, he quit Tokyo and hurried to California, hurried to the social contacts Lloyd possessed, if not offered, and to Barnsdall's clique. Lloyd's first wife was the actress Elaine Hyman who used the stage name Kira Markham. Their social and professional circles were with Chicago and New York theater tribes, with other artists and artistes and, of course, after 1919 with Hollywood people.

The "sentimental" interlude of the "Romance holiday" in Japan (his words), was a culmination that began in 1913 with the Imperial Hotel. Wright told the Amsterdam architect Hendrik Petrus Berlage in 1922 that, "Yes, you are right. I have been romancing—engaged upon a great Oriental Symphony—when my own people should have kept me at home busy with their own characteristic industrial problems...". It was the American people who failed him; they were at fault. In self-pity he openly acknowledged hurtful experiences under society's cruel scrutiny and explained it in this manner:

> [D]ear Dr. Berlage, I am branded as an "Artist" architect, and so under suspicion by my countrymen —and especially as I have been an "insurgent" in private life as well as in my work; and my hair is not short nor my clothes so utterly conventional as to inspire confidence in the breast of the good American Business Man that I am a good "business proposition". It cost more to employ me....

He then acknowledged that the cost was not just of money. To hire him could be a difficult social burden upon a client and those involved in a building's construction:

> it's a matter [of] imposing independent thought and action and some pains upon the man who employs me. It is difficult to achieve "the greater end", and part of that difficulty is the client's. So I have had to go where opportunity led me, and I have had very little choice.

Wright did not mention society's reaction to stories about abandoning wife and children in 1909 for a married lover, and after her murder taking on a new lover, and months after her collapse taking on another. Yet he did infer those socially unacceptable actions, and others, saying he was spurned by society and could only obtained independent minded clients. And so began his migrations to Tokyo, Los Angles, and occasionally New York City

As to his just completed Tokyo experiences that had consumed so much energy and tested his oriental hosts, in his personal elliptical prose Wright told Berlage that, "The Japanese seemed enthusiastic and grateful but were being subjected to criticisms of every kind and nationality—the British being the only ones adverse except countrymen of my own".[69] Indeed, generally speaking his American friends and clients between 1916 and 1932 tended to be thoroughly independent and on the margins of acceptable society: left wingers, communists, anarchists, free sexers, progressives and old Progressives, and (of all horrors) poets, theater people and others Artists.[70] More or less befitting their passions and life styles, their Wright houses were also atypical and special. Four of Wright's initial clients in Southern California were involved actively in or supportively with Barnsdall's leftist art and theater circle and/or other art and literary coteries like the Hollywood Art Association. So too were Lloyd and his wife, Neutra and his wife and especially Schindler and his wife. The involvement of Schindler, Lloyd and Neutra in literary and art circles and with the Hollywood film industry was crucial to the success of their professional careers.[71] Wright Senior was shunned.

As a further embarrassment, some of Wright's clients transferred allegiance, so to speak, and hired former employees and Lloyd. We can recall the Lowes preferred Schindler, and later so did the Freemans, Millard, and Barnsdall. The Ennises found Wright a *persona non grata*. Between 1924 and 1927, when the city accepted acquisition of the Olive Hill campus, Schindler produced for Barnsdall, twelve designs for changes and improvements. Wright Senior was totally ignored. She commissioned Schindler to design a new mansion, the "Translucent house, aspects of which recall Wright's Coonley, Barnsdall and Ennis houses".[72] Lloyd designed a major addition to Millard's house using father's textile blocks. It is true that around 1929 Mrs. Millard, with some trepidation, asked Wright for architectural designs that would add value to her house and property in such a way that would insure their preservation; but they came to naught. Schindler produced a few minor internal changes for the Freemans. Neutra attempted in influence political affairs as related to Olive Hill but that was in 1931 when Lloyd was more favorably involved with the Barnsdall property.

There was Doheny's rejection, Johnson's withdrawal, the collapse of the Tahoe and Nakoma projects, and what else? Wright Senior was stymied. His troubled mind and quick temper, ("I have a caustic way" he once confessed), is patently obvious in a very long and detailed letter he wrote to Barnsdall, prepared in June 1921.[73] From Chicago and well before those major reversals and refusals, he wrote to Louis Sullivan in November 1922 that he had "a secret" for "Leiber Meister" to keep: "I am extremely hard up, [Wright said,] and not a job in sight in the world. My 'selling' campaigns have failed".[74] Yet, with exception of Tokyo his campaigns were performed after that 1922 letter. Other than for himself or family, Wright's independent commissions after the Ennises in 1923 were in 1928 (a cabin), then 1933 a house for the Willeys in Minneapolis, Minnesota. Sullivan was also in dire straights. His architectural practice was moribund and he needed friendly cash on which to live. Independent of one another and among other Chicago people, Wright's son John and Schindler had financially helped Sullivan, so too had Wright but not on that 1922 occasion. Wright's "selling" campaign will be outlined shortly.

CONCRETE ADOBE

On arrival in Los Angeles in 1922, Romance was discarded in a "hunger for reality", Wright said, a need for a rational architecture through construction and with elemental forms.[75] Satisfaction came by a re-examination of the theory of "pure design" that had stimulated designs for Unity Temple and the Larkin Building and

Beye Boathouse project, the exemplars,[76] and to embracing the severe reductive and subtle regional qualities he observed in Gill's buildings. The influence of Gill at this delicate moment in Wright's—and Schindler's and Lloyd's—career cannot be over emphasized. A Gill-like simplicity of parallelepiped forms, of interlocked or juxtaposed boxes, is clear in Wright's unique textile block houses for Millard, Storer, Freeman and Ennis. Their simple volumes defined interior spaces and expressed functional disposition: Like Gill's it was a truly functional architecture.

The Ennis house, therefore, is a mix of Gillian severity and molded concrete adobe. The battered and stepped (please note) walls and winged Palladian window may have been concessions to the Ennis' enthusiasm for things Mexican and Indian.[77] But the floor plan and massing are a further exploration of what Wright described as "articulation", or what in 1908 he referred to as the "individuality of various functions". That is, the internal spaces and three dimensional masses of them are individuated. This form-follows-function notion was first explored for office additions in 1898 to his home at Oak Park, Illinois, for buildings such as Larkin and Unity Temple, for the McCormick house project, for Beye Boathouse project, and so on.[78]

The Ennis dining area is defined internally by post and wall and externally by a volume encrusted overall with block patterns that differ for each distinct space and volume of living, sleeping, passageways, and separated servants' quarters. The respected architectural journalist Douglas Haskell correctly noted in 1928 that "the outside mass declares the inside room".[79] A winged relief formed of concrete tiles on the exterior is above a great dining window that overlooks the roofs and streets of Hollywood and other neighborhoods of Los Angeles and to the sea.

Functional distinctions were only slightly less obvious for the Storer and Freeman houses, Millard's was more compact yet parts are also individuated: each plan parti makes that clear. Additionally, the only difference in external *appearance* between those houses, Gill's, and the Indians' high desert buildings was the textured walls. Patterned blocks and tiles on the interior created a slightly claustrophobic but Southwestern richness, slightly more so for the Freeman house. Most inside block faces for Freeman and Millard were plain, the rooms cave-like. After 45 years in his home, Samuel Freeman remained satisfied:

> *Now you could take the [living] room and strip everything [furniture, etc.] out of it, and you wouldn't feel that you were in an empty room. This room itself is a piece of sculpture. It has a life of its own. . . . It has broken planes, different heights . . . the whole thing like music . . . and I'm never bored with the room. . . . [It's] always exciting to me. It's almost alive, it's in motion.*[80]

(The Griffins provided practicalities for the Wrights' initial blocks, but the Griffins' and Lloyd's buildings never—never—attained the refinement, the color, the elegance, the magic of Wright's. Lloyd and the Griffins were ordinary architects acting in unusual situations that now attract our attention. In city or community planning, however, the Griffins were superior: Lloyd naive; Wright Senior lacked perspicuity.[81])

It is interesting to speculate on the level of Schindler's influence on the Olive Hill buildings. Schindler studied at the Imperial Technical Institute before receiving degrees in architecture and engineering from the Academy of Arts in 1913. He had left for the U.S. almost immediately upon receiving degrees so he had not met Griffin when the latter traveled to Vienna in 1914 to enlist Wagner as a jury member. While living in Chicago, Schindler had made a pilgrimage in 1915 to New Mexico, Arizona, and Southern California where he saw Gill's buildings. He photographed and sketched native American buildings and even made some designs based on Pueblo architecture. The general three dimensional overall outline of his unrealized T.P. Martin house of 1915 for a site at Taos, New Mexico, its soft-corner adobe character, the parti of the U-shaped floor plan and functional disposition, and especially aspects of the central court as portrayed in perspective, anticipate almost wholly the Barnsdall house design begun four years later.[82] See Figs 7.7-7.8.

Schindler started with Wright in February 1918 and in early 1919 the Barnsdall house on Olive Hill was at the conceptual stage. While it is sometimes said that Wright began designing the house in 1915, this is difficult to believe. He seldom designed in the theoretical abstract for a client except in the early 1920s. Barnsdall selected the Olive Hill site early in 1917 and purchased it that June. It was while Wright was in Japan that Schindler produced most of the preliminary and final drawings, all dated 1919 or after, assisted in later stages by sons John in Chicago and Lloyd in Los Angeles. Wright Senior was mentally and emotionally consumed with construction of the Tokyo hotel to a point where possible distractions were put aside. However, the floor plan is schematically also rather typical of Wright's large houses (sans the "garden courtyard") beginning with Coonley's at Riverside, Illinois (1906-08), and projects for Shaw in 1906 and McCormick in 1907. As the Martin house floor plan implies, Schindler must have been familiar with them on visits, or as part of office lore or as published in 1910: they were not typical American practice.

One can only speculate on the degree of influence Lloyd and Schindler—at times in discussion or by their own designs—may have had on Wright as to the theoretical reasonableness of a more obvious regional response. Remembering that Gill, Schindler and his wife, and Lloyd were close friends, and that Gill's work was greatly admired;[83] remembering Wright's buildings in the blue green northern climes, or the non-regional aberration of his Stewart house of 1909 at Montecito near Santa Barbara (that was before his 1913 and 1915 trips through the Southwest to San Diego and Los Angeles and before Schindler's presence in the office); remembering too Lloyd's experience with Gill, and Gill with Wright; remembering that Wright's designs before 1919 do not even hint at the Monolith Homes or Lowes house projects or the five Los Angeles houses; while beaten socially Wright surely listened to those close to him and was quite able to adapt artistically as a necessity and regenerate a social and professional life.[84]

1922: work in the Los Angeles office was left to Schindler and Lloyd. For the Olive Hill buildings, Lloyd was responsible for landscape (and he prepared a

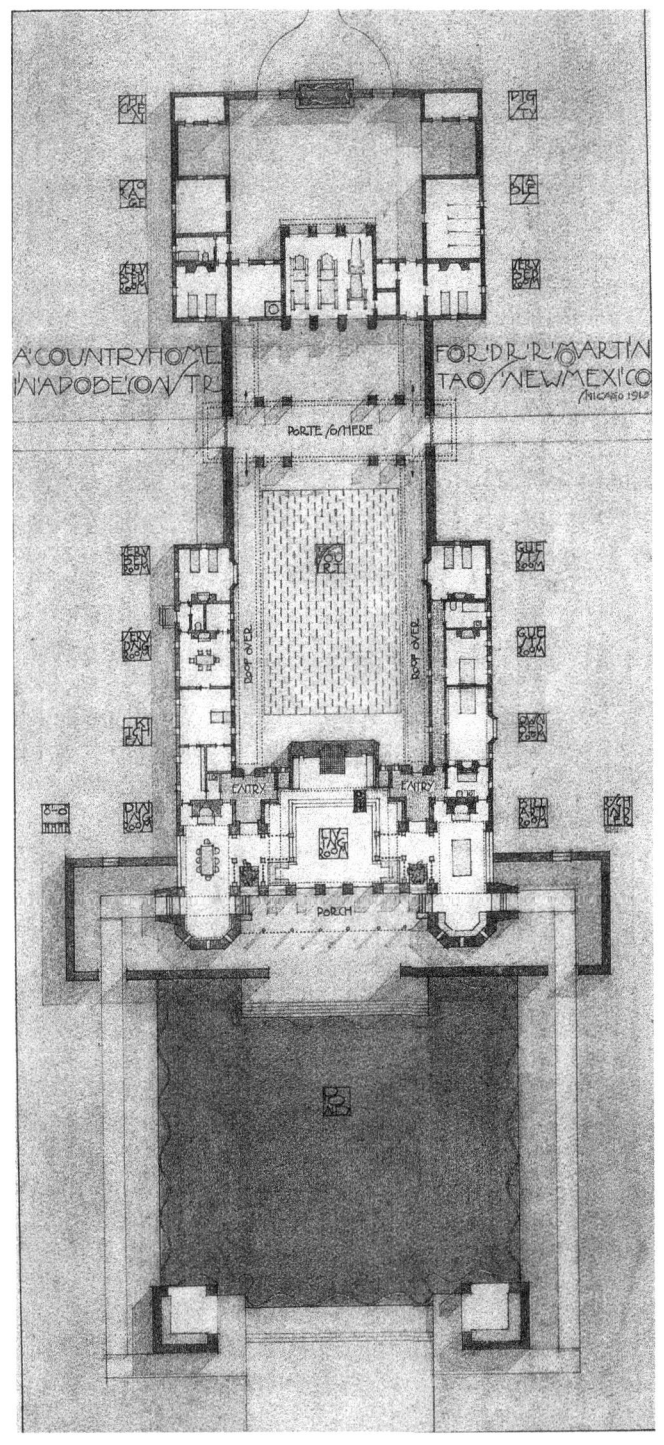

Fig. 7.7 T.P. Martin house for Taos in New Mexico, R.M. Schindler, Architect, a project of 1915

Fig. 7.8 Martin house, perspective drawing by Schindler of internal "court" looking toward living room

master plan that was not followed) and Schindler for the development of building designs and construction details: both supervised construction. Residences B (the Director's House) had a similar parti to the Ennis house but with two faces: Wright's to the street, Schindler's to the campus.[85] The design of Residence A is usually credited more to Schindler than his boss, and correctly so. The late historian and theoretician Colin Rowe, for example, found it was, he insisted, "*so very* Wagnerschule",[86] a school Schindler did not altogether abandon. That influence is not noticed for the textile block houses because, perhaps, by late 1922 Schindler had left Wright at the moment when "the master" decided to reside permanently in America and study Griffins' knitlock and Lloyd's knit-block.

NOTES

Epigraph: Kamerling (1993), title page, p. 124, as taken from Irving J. Gill, "The Home of the Future. The New Architecture of the West. Small Homes for a Great Country", *Craftsman* (New York), 30(May 1916), pp. 140-151, 220.

1 Horatio Greenough, *Form and Function. Remarks on Art*, Harold A. Small editor, (Berkeley: University of California Press, 1947), introduction by Erle Loran, pp. xviii-ix. See also Horatio Greenough, "Form and Function" in Lewis Mumford, editor, *Roots of Contemporary American Architecture* (New York: Reinhold, 1952), pp. 32-56.

2 Gebhard (1990), p. 148; see also David Gebhard "The Spanish Colonial Revival in Southern California (1895-1930, *JSAH*, 26(May 1967), pp. 131-147, also on Gill, Lloyd and the Ennis house.

3 As quoted in Kamerling (1992), pp. 125-126, see note 1 above. Cf. Katie Nicole Moss, "Constructing a modern Vienna: the architecture and cultural criticism of Adolf Loos", M.A. thesis, Graduate School, University of Oregon, Eugene.

4 Hebbard was a graduate of Cornell University (1887) and enroute to San Diego worked for architects in Chicago and Los Angeles; partnership with Gill began in 1898.

5 Mead was trained as an architect's apprentice and at the Pennsylvania Museum and School of Industrial Art 1889/90 and worked in Penn's town until 1901. In 1903 he showed up in San Diego. Beginning 1908 he travelled to the southwest of Arizona and New Mexico and became an activist for the displaced Mojaves-Apache and then supervisor of their new reservation. In 1912 he formed a partnership with architect Richard S. Requa, also trained by Gill. Mead began a solo practice in 1920.

6 Kammerling (1993), p. 37; Kathleen Flanigan, "William Sterling Hebbard: Consummate San Diego Architect", *San Diego Historical Society Quarterly*, 33(Winter 1987) pp. 9-33.

7 Kamerling (1993), pp. 125-128.

8 Wright (1932), p. 239.

9 Wright (1932), p. 240, emphasis FLW. See also Lloyd Wright and Wright Senior correspondence early April through June 1932 in Wright Archive and LWPapers.

10 Gutheim (1941), p. 124, a draft and not as published.

11 Architectural hybridization is a continuing study by this author, but see an early case study, Donald Leslie Johnson, "Vernacular Architecture: a Theory of Hybridization", *Historic Environment* (Melbourne), 3(2, 1983).

12 Hines (2000), p. 71.

13 All quotes Hines (2000), p. 73; see also Roorback (1921).

14 As quoted in Marvin Rand, *Irving J. Gill: architect 1870-1936* (Layton, Utah: Gibbs SMith, 2006), p. 46.

15 Hines (2000), p. 73

16 Hines (2000), p. 73.

17 As quoted in Esther McCoy, *Irving Gill, 1870-1936* (Los Angeles: County Museum, 1958), p. 17.

18 Sally Kitt Chappell and Anne Van Zanten, *Barry Byrne and John Lloyd Wright. Architecture & Design* (Chicago: Chicago Historical Society, 1982), Golden West Hotel p. 46; Barbara Stodola, et al, *John Lloyd Wright and colleagues: Indiana works* (Michigan

City: John G. Blank Center for the Arts, 1999), pp. 10-13, 39; Richard Guy Wilson, "Theme of Continuity", in idem and Sidney K. Robinson, ed., *Modern Architecture in America* (Ames: Iowa State University Press, 1991); "Earliest work of John Lloyd Wright", *Prairie School Review* (Chicago), 7(1, 1970), pp. 16-19.

19 Kammerling (1993), p. 42.

20 Biographical outlines also based on David Gebhard in Macmillan; Sarah J. Schaffer, " 'A significant sentence upon the earth'; Irving J. Gill, Progressive Architect", *San Diego Historical Society Quarterly, "Part I", 43(Fall 1997), "Part II", 43(Winter 1998);* Gebhard/ Von Breton (1971); Gebhard (1971), chapter 3; Thomas S. Hines, "The blessing and the curse", in Weintraub (1998), pp. 11-27; J. Wright (1946), pp. 129-133; "Irving J. Gill (1870-1936)" at <www.sandiegohistory.org/bio/gill/gill.htm>, accessed 31 October 2007. On Gill and Schindler, McCoy/Makinson (1960); Smith/Darling (2001); Smith (1992), p. 80.

21 J.Wright (1946), pp. 131-135. For background see Harold Kirker, *California's Architectural Frontier . . . Nineteenth Century* (New York, 1960, reissued 1970, reprint Salt Lake City: Peregrine Smith, 1986); idem, "California Architecture and Its Relaltion to Contemporary Trends in Europe and America", *California Historical Society Quarterly*, 51(Winter 1972); Weitze (1984); and David Gebhard and Robert Winter, *Architecture in Los Angeles* (Salt Lake City: Peregrine Smith, 1985).

22 Wright (1946), p. 134; on Gill's influence on former colleagues, McCoy/Makinson (1960), and see the Gill-inspired and mature double house for O.S. Floren, Hollywood (1922), among a few others of the period, Gebhard (1971), pp. 65-75; and cf. David Gebhard, "Irving J. Gill", in Robert Winter, ed., *Toward a Simpler Way of Life. The Arts & Crafts Architects of California* (Berkeley: University of California Press, 1997), pp. 201-208, 229-239.

23 As quoted in Hines (2000), p. 156.

24 FLW to L.Wright, 1 June 1932, copy Wright Archive.

25 Wright to Robinson (at *Fortune* magazine), 14 May 1932, p. 3, copy Wright Archive.

26 Cf. J. Glass, "Realism or idealism? Tilt-up construction pioneers", *Architectural Science Review* (Oxford), 44(1, 2001), pp. 53-59. The Tilt-Up Concrete Association initiated the Irving Gill Distinguished Architect Award in 2003 to recognize TCA members committed to persuing architectural design excellence. Gill used the Aiken tilt-up system, then purchased patent rights and formed his Concrete Building and Investment Company.

27 Mitchell Schwarzer, Irving Gill in *Encyclopedia*, vol. 2, p. 504.

28 As quoted in Kamerling (1993), p. 57.

29 Hines (2000), p. 13.

30 Wright (1932), p. 227.

31 FLW, *Genius and the Mobocracy* (New York: Duell, Sloan & Pearce, 1949), p. 52.

32 Hines (2000), p. 132.

33 Christopher Long, "The Origins and Context of Adolf Loos's 'Ornament and Crime,'" JSAH, 68(June 2009), p. 217.

34 Leonard K. Eaton, *American Architecture Comes of Age* (Cambridge, Massachusetts: MIT Press, 1972), p. 139; cf. Donald Langmead, "Adolf Loos", Johnson/Langmead (1997), pp. 94-98; Janet Stewart, *Fashioning Vienna: Adolf Loos's Cultural Criticism* (New York: Routledge, 2000).

35 Wright (1910), plate 43a/b: Munz/Künstler (1966), passim; Mitchell Schwarzer, "Adolf Loos", *Encyclopedia*, vol. 2, pp. 720-723.

36 Munz/Künstler (1966), pp. 130-131, figures 131-133, the Scheu house was published well after 1918. Loos also mentioned a likability for the then-popular Japanese vernacular designs.

37 Munz/Künstler (1966), pp. 201-204.

38 Hoffmann (1992), p. 37; see also Levine (1996), p. 454,n.70.

39 Wright (1925), p. 59.

40 Wright (1925), pp. 56-59, the signature on one of the drawings is "J Ll Wright Taliesin Dec 7[?] 22". Lowes built a house in 1923 to a design by Schindler and independent of Wright.

41 FLW ms. for a lecture (given at Hollyhock?) 1931, copy LWPapers, *similar* to Wright (1932), pp. 227-228.

42 As quoted in Art Ronnie, "Hollyhock—The Wright House", *Westways* (Los Angeles), 66(November, 1974), pp. 18-22, based on interviews with Lloyd Wright. Apparently similar comments were made to Levine (1996), p. 455,n75. See also "FLW's Hollyhock House", circular, Los Angeles Municipal Art Department, n.d. but ca.1931; Smith (1992), p. 27n; Torrance (1982), p. 146, 173; Norman M. Karasick, "Art, Politics, and Hollyhock House", Master of Arts thesis (California State University, Domingues Hills, 1982), p. 12.

Roma*nza* is Italian for romance, a term not architectural but used with vague musical significance that FLW understood, but he tried to connect them. "Romance" in Spanish is *romance*.

43 Drawings by Schindler, Smith (1992), figures 71-79; cf. Mark L. Brack, "Domestic Architecture in Hispanic California ... Reconsidered", *Perspectives in Vernacular Architecture, IV*, Thomas Carter & Bernard L. Herman, editors, (Colombia: University of Missouri Press, 1991), pp. 163-173.

44 Levine (1996), p. 136-156ff.

45 Wright sent illustrations to Wijdeveld first in 1920, again in 1923-24, then again for a series of seven articles in 1925 that were brought together in book form in Wijdeveld (1925), pp. 131-157. See Langmead/Johnson (2000), chapter 6. Photographs of pueblos were widely published in a variety of books and magazines. One probably seen by Wright was Vere O. Wallingford, "A Type of Original American Architecture", *ARecord*, 19(June 1906), pp. 467-469.

46 I have used the word "canted" as in a slanted or tilted position. On one of the Barnsdall drawings, sheet 11 dated 20 August 1919, the word "chamber" is used, but the word refers to a compartment or cavity like a room. "Camber" refers to an upward curve; "chamfer" refers to a corner cut off as in "bevel". I've also used the word "battered" and that refers to a surface that slopes inward toward the top. The word "attic" has been used from classical language as the upper most story of a building, or a surface area formed by a sloping roof.

47 A useful example is "The Dance of the Great Knife" at Zuni Pueblo in New Mexico illustrated in Cushing (1882).

48 The balustrade attic is detailed in Patterson (1994), p. 192, based on construction drawings. Located just beyond the open end of the U-shaped courtyard is a tiny garden hemicycle, a place for contemplation, often misconstrued as seating for a theater in the round.

49 As quoted in Friedman (1992), p. 251; and Friedman (1998), p. 51 and passim, where she also ineffectively makes a case for Pueblo influence. The role of the courtyard and roofs was surmised by this author in the 1960s, confirmed during visits to Hollyhock and then by verbal and published evidence. See also Smith (1992), pp. 79-80, 217,n5.

50 H.P. Berlage, "FLW", in Wijdeveld (1925), but the letter is not extant, with thanks to the late Dr Donald Langmead. See Langmead/Johnson (2000), chapter 7.

51 Mark Gelernter and Virginia Dubrucq, "Regionalism", *Encyclopedia*, vol. 3, p. 1089; Frank N. Owings Jr., "FLW and the Regionalists: Visions for America", *FLWQ*, 14(Winter 2003), pp. 4-21, looks at Wright's relationship with painters of the 1920s and 1930s who, like that group of writers around 1900, practiced an art identified with a geographic region.

52 Glade Sperry, Jr. "Pueblo Images in Contemporary Regional Architecture: Primal Needs, Transcendent Visions", in Marcovich (1990), p. 289.

53 Thomas (2000), p. 138.

54 Arnold Berke, *Mary Colter. Architect of the Southwest* (New York: Princeton Architectural Press, 2002), p. 9; various web sites January-March 2012..

55 Often reprinted, see Cushing (1882), the Palmer Lake, Colorado (1967) edition contains original illustrations and a short biography of Cushing by Oakah L. Jones, Jr. Cushing's study of the Arizona Havasupai was published in two issues of *Atlantic Monthly* in 1892, and his book *Zuni Folk Tales* was published in 1901. The controversial Cushing began with the Bureau of Ethnology when age 18 and remained until a premature death at age 43. See also Thomas (2000), chapter 8; and Brian Fagan, *Elusive Treasure* (New York: Scribner's, 1977), chapter 12; idem, *The great journey* (London: Thames & Hudson, 1987).

56 *California History*, vol. 71 (Fall 1992) is a useful special issue devoted to local Indians.

57 Thomas (2000), pp. 60-78. In 1491 there were about 10 million indigenous people in North America. In 1900 the aboriginal population was down to around 250,000, Francis Jennings, *The Founders of America from the Earliest Migrations to the Present* (New York: W.W. Norton, 1993), throughout; Thomas (2000), part 1.

58 Observation of Gebhard/Von Breton(1972), after extensive interviews with Lloyd Wright.

59 A good photographic view of L. Wright's Hollywood Bowl of 1927 is Torrance (1982), p. 128.

60 The interesting history of these projects is told in Mary Jane Hamilton, "The Nakoma Country Club", and "Nakoma Memorial Gateway", in Sprague (1990), pp. 77-88. See also De Long (1996), pp. 47-66; Riley (1994), plates 172-174; Levine (1996), pp. 155-156; Alofsin (1993), pp. 280-294. A building very similar to those for the Nakoma country club was Wright's design of the Davis house of 1950, Patterson (1994), pp. 136, 145. Good illustrations of Director's House and Residence B are in Smith (1992), chapter 6, pp. 89-98, chapter 14.

61 Two designs were prepared in 1924 and 1925 for Mrs Edna Browning Gladney, philanthropist and childrens' social activist, of Sherman, Texas; De Long (1996), pp. 191, 193; for 1925 design see Alberto Izzo and Camillo Gubitosi, *FLW Three Quarters of a Century of Drawings* (London: Academy, 1976), no. 71; Arthur Drexler, *The Drawings of FLW* (New York: Museum of Modern Art, 1962), no. 105.

62 Eric Mendelsohn, "A visit with Wright", letter cited in H. Allen Brooks, *Writings on Wright* (Cambridge, Massachusetts: MIT Press, 1983), pp. 7-8; see also Thomas (2000), pp. 139-140.

63 Thomas (2000), p. 184.

64 L.Wright to Alice Robinson at *Fortune* magazine, 14 May 1953, p. 4, copy LW Papers; and FLW to L. Wright, 1 June 1932, copy Wright Archive.

Schindler mentioned that for the Barnsdall campus Wright introduced "Mayan motifs" that would "give themselves [the buildings] local roots", Smith (1992), p. 43, but he did not identify the motifs. He also said the buildings had roots in ancient Greece. No doubt the two comments were based on those by Wright.

65 FLW, "1928: In the Cause of Architecture: Purely Personal", ms. published (and edited?) in Gutheim, (1941), p. 130.

66 Wright (1910), n.p. but last line of text.

67 Alofsin (1993), throughout.

68 On Wright's travels in the Orient see Meech (2001), chapter 1-5; Alofsin (1993), pp. 309-310; Smith (1985), pp. 296-310. On Wright's career in 1910s and 1920s see relevant chapters in Wright (1932); De Long (1996); Twombly (1979); Secrest (1992); Gill (1987).

69 FLW to Berlage in Amsterdam, 22 November 1922, copy Wright Archive; Pfeiffer (1984), pp. 54-55. The importance of Berlage's visit to the U.S. and especially Chicago and Minneapolis in 1922 is explained in Christopher Vernon, "Berlage in America: The Prairie School as 'The New American Architecture'", in *The New Movement in the Netherlands 1924-1936*, Jan Molema, editor (Rotterdam: 010 Publishers, 1996), pp. 131-151; and Langmead/Johnson (2000), chapter 1.

70 Friedman (1992), pp. 242-244.

71 A further hurt was caused by a young Antonin Raymond who, as an employee, had traveled with Wright from Chicago to Tokyo in December 1919, quit in early 1921 and immediately acquired clients to begin a long, successful career in Japan; see Raymond, *An Autobiography* (Rutland/Tokyo: Charles E. Tuttle, 1970).

72 Smith/Darling (2001), p. 27.

73 Smith (1985), Appendix 2, pp. 211-213.

74 FLW to Sullivan, 30 November, 1922, copy Wright Archive. Client A.M. Johnson was a Chicago businessman with an interest, we're told, in the geology of Death Valley.

75 Wright (1932), pp. 235-236.

76 Discussed in detail in my forthcoming book, "Education and FLW".

77 Early in the design process the walls for the Ennis house were to be stepped rather than canted as built, see *Treasures*, p. 104; Riley (1994), p. 208 See also Alofsin (1993), p. 296. Lloyd Wright supervised construction of the Barnsdall, Millard, Storer, Freeman and Ennis houses.

It wasn't until the 1930s that Wright took note of his contemporaries in Mexico; see Keith Eggener, "Towards an organic architecture in Mexico", in Alofsin (1999), pp. 166-183.

78 Wright (1908), p. 160; Wright (1928), pp. 56-57; see also Rebori (1927), pp. 449-454; see also my forthcoming on "Education and FLW".

79 Haskell (1928), p. 11.

80 Meehan (1991), pp. 62-63. For Wright's buildings constructed in California see David Gebhard and Scot Zimmerman, *Romanza. The California Architecture of FLW* (San Francisco: Chronicle, 1998); also Dunham (1994); Pfeiffer (14-24); Pfeiffer (24-36).

81 Johnson (2004), throughout

82 Smith/Darling (2001), pp. 20,22,182,189, illustrations pp. 184-188; Gebhard (1971), pp. 278-279; McCoy (1979), p. 26; Albert Narath, "Modernism in mud: R.M. Schindler, the Taos Pueblo and a 'Country Home in Adobe Construction'", *The Journal of Architecture* (London), 13(August 2008), pp. 407-426.

83 Hines (2000), pp. 231-232.

84 The following provide valuable historical information and insights for the period 1914 to around 1924. The letters are accompanied by a good contemporary context; correspondents and their subjects are wide ranging: McCoy (1979); Dione Neutra, compiler & translator, *Richard Neutra Promise and Fulfillment 1919-1932. Selections from the Letters and Diaries of Richard and Dione Neutra* (Carbondale: Southern Illinois University Press, 1986).

85 Illustrated in Wijdeveld (1925), pp. 148-151, drawings by Schindler and L. Wright; Smith (1992), pp. 158-159, 173-179. Residence B was demolished in 1949.

86 As Rowe said to Hines (1995), p. 474. Other similar observations are Reyner Banham in discussion with this author in 1989; Reyner Banham, *Los Angeles. The Architecture of Four Ecologies* (Harmondsworth: Penguin, 1971), pp. 179-191; August J. Sarnitz, "Integrity and Ambiguity", and Barbara Giella, "Buena Shore Club", in Lionel March and Judith Sheine, editor., *RMSchindler* (London: Academy Editions, 1993), pp. 77-86, 43-44; Anthony Alofsin, "FLW and Modernism", in Riley (1994), p. 39; Gebhard (1971), pp. 41-42.

8

Closure . . . Schindler and Resurgence

> The world to come must be composed of what is past. No other material is at hand.
> — Cormac McCarthy, 1998

> We walk backward into the future.
> —Hopi legend

> Architecture is beginning, always beginning . . . . It is something that has to be made afresh all the time, as life, as opportunity, as growth changes.
> — Frank Lloyd Wright, 1940

Lloyd Wright continued his landscape design activities and in 1927 formally opened an office in Los Angeles to also practice architecture. When he specified concrete blocks they were of his own design, almost always non-structural and highly textured as ornament. Later he was joined by his architect son Eric.

Overt action against the avant garde and the reinstitution of historicism, and a faltering progressivism in the 1920s affected all of the architects under discussion, especially Gill and Wright Senior, less so Lloyd, Schindler and Neutra who managed to obtain the odd job. After construction of a few buildings designed by Gill the promoters and workers of Torrance, a new—supposedly model—industrial suburb south of Los Angeles, rejected his building designs; then the entire housing scheme collapsed. It was not long before Gill abandoned Los Angeles and by the force of economic circumstances all but retired by 1933. But Gill always remained close to Schindler and the Wrights. He died in 1936.

In October 1921 Schindler quit Wright. His first independent act was to build his own house that he had been designing for a number of months. It was to be beside Kings Road in West Hollywood, a house without the slightest hint of Wrightian idiosyncracies except, perhaps, wood work stained by ferric oxide. Inspired by trips to New Mexico tribal lands and later a vacation in Yosemite National Park in 1921, in November of that year Schindler presented a house plan of two linked single rooms or "studios", one for the Schindlers, one for friends Clyde and Marian Chase.

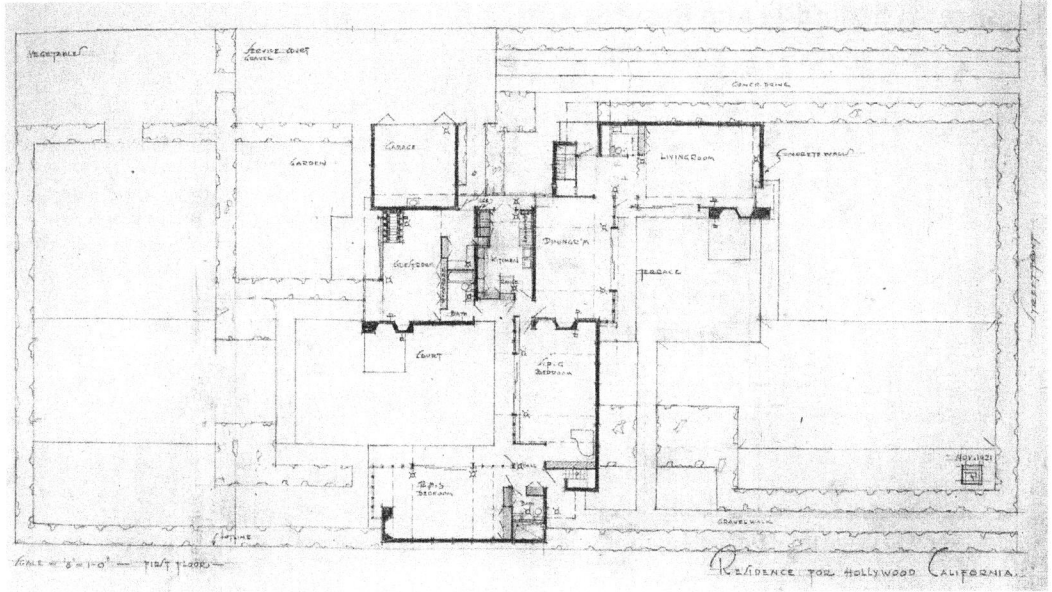

Fig. 8.1 Architect R.M. Schindler's own Kings Road House in West Hollywood, 1921-1922

Each studio was served by a single kitchen-laundry-service area. See Fig. 8.1. There were no rooms for living, dining or beds. Two roof-top sleeping areas had canvas sides. There was also a small fully equipped guest studio.[1]

Schindler was familiar with the concrete tilt-slab system employed by Gill. Rather than tilting up an entire wall as Gill had done, Schindler used tilt-up sections of wall that were manageable by two men using a relatively small tripod lifter. It was a unique system and, oddly, not again used by him or his colleagues.[2]

Complete in May 1922 the house together with Gill's Horatio West Court of three years earlier were giant leaps forward in defining a new modernism. They set out with utter simplicity the functional plan while Schindler introduced a horizontal flow of space and floor plane (polished concrete) to exterior court and garden through a glass wall that could be pulled aside, the interior thereby intimate with the exterior that a mild climate allowed. As had Gill for the past fifteen years, Schindler reduced internal aesthetics by eliminating ornament except as it might be seen by the exposed wood roof structure and ceiling, detailing cabinetry, windows and glass walls, by sunlight and shade, and at Kings Road by copper fire place hoods. Schindler's exterior walls, Fig. 8.2, were concrete (left natural inside and out) and glass; the ceiling and roof structure darkly stained redwood. There was no formal plan or elevational axes: the elevations responded directly to the plan and materials naturally. Except for the brilliantly conceived and executed Dr James Eads How house of 1924-1925,[3] Schindler did not again design with such theoretical clarity until the early 1930s. Neutra did not catch up to Schindler's spatial, structural and architectonic concept until 1927, Wright in the mid-1930s. Europe's sever northern climate was a critical determinant in maintaining internal discrete chambers and the resulting three-dimensional forms. Southern California's mild sunny climate inspired the Austrian emigrants' architectural responses.

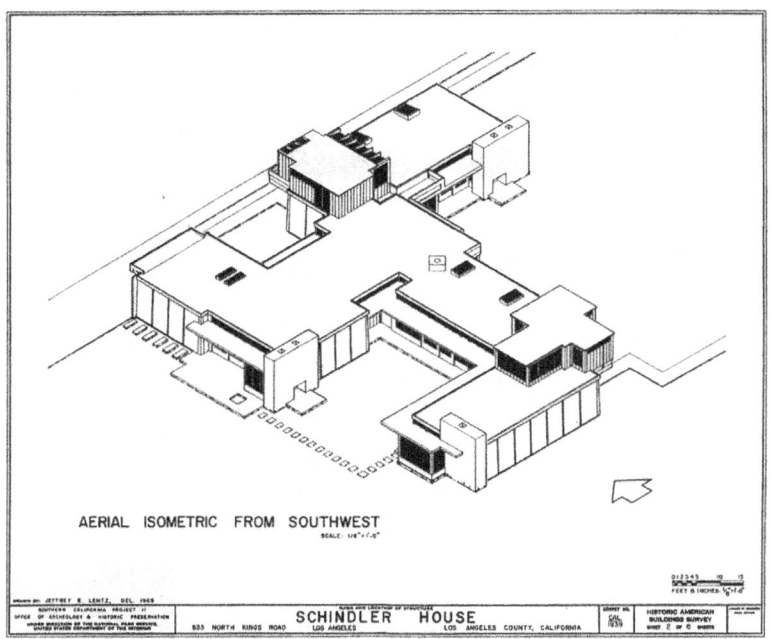

Fig. 8.2
Schindler's Kings Road House, "aerial isometric" of 1969

One of Schindler's first clients was Dr. W. Llewellyn Lloyd. While Kings Road was an architectonic experiment for personal pleasure, Lloyd's program of "bungalow apartments of distinction" to be called El Pueblo Ribera Court, was a housing investment in fashionable La Jolla just north of San Diego, see Figs 8.3 and 8.4. Schindler outlined his design response of 1923:

> *I propose to treat the whole in true [south] California style, the middle of the house being the garden, the rooms spreading wide into it, the floors of concrete, close to the ground. The roof is to be used as a porch, either for living or for sleeping, and should be one of the features of the place, with its ocean view. . . . The [prototypical] unit is planned in such a way that it can be closely joined and combined with other units, [Fig. 8.3] without sacrificing privacy of rooms, garden, or roof.*[4]

Kings Road House was an experiment in tilt-up construction but Ribera, was a trial of the moveable wood lift- or slip-form system. Engineers had exploited it to a very high degree for the great tall grain silos on mid-Western prairies at least twenty years earlier and Lloyd Wright had used it. Schindler's description and deployment of Ribera Court units on site were based on the parti of his own house in the manner of relating interior to exterior and in gaining exterior visual privacy by extended blank back walls. His aesthetic response was to allow the building materials (a la pueblo?) to stand unadulterated by paint or plaster while rooms were nicely articulated. As with Gill's Horatio West Court in 1919, Ribera too was a truly objective functional response. Unfortunately Dr. Lloyd's housing scheme was not financially successful. In April 1925 he confessed promotional difficulties to Schindler, saying that "Everybody praises the places to the skies, but only a few

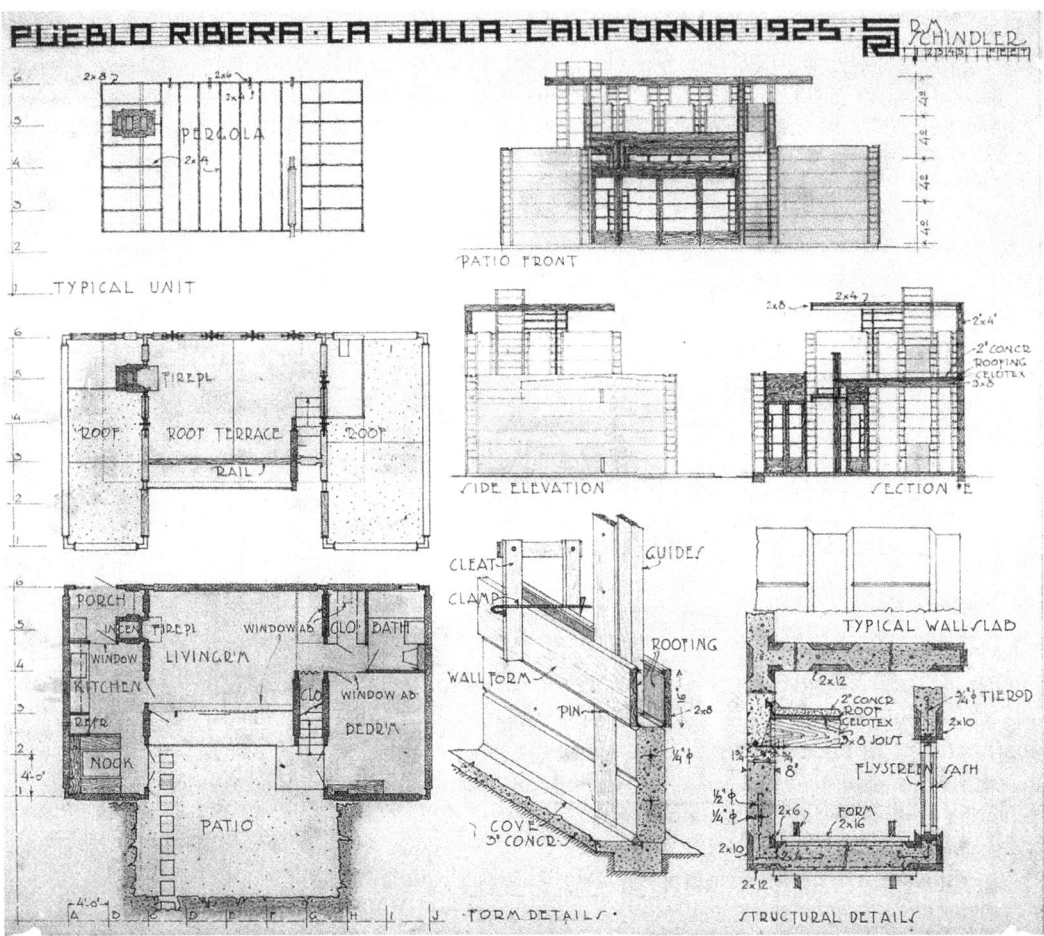

Fig. 8.3 El Pueblo Ribera Court in La Jolla, California, R.M. Schindler, Architect, 1923-1924

have the courage, apparently, to live in them".[5] Ribera Court has survived a fire and remodeling and in 1977 was declared a San Diego Historical Site.

From early 1925 to 1927 and while Neutra and family lived in the Chase apartment (the Chases then in Florida) the two Austrian emigrant architects occasionally worked together but not in a functional or legal sense. After witnessing Wright's bungling, neither was game enough to design or employ concrete blocks as structure or ornament. (Yet Schindler did begin to detail "slab block construction" of concrete wall tiles as late as October 1939;[6] but they remained paper ideas.) After perloining a Schindler client the ebulliente Neutra rightly gained high acclaim internationally beginning instantly with publication of the Lovell House of 1927-1929. In the post-1925 rush of North Americans to eulogize and emulate European modernism, the radical designs of the other two architects were forgotten. Now we can fully appreciate Morrill's and Gill's revolutionary architecture 1907 to 1920 and the precendent of Schindler's Kings Road house, El Pueblo Ribera and How house. Exactly how their revolutionist works came to be valued in Europe is to be studied.

Fig. 8.4 El Pueblo Ribera, 1923-1924, photograph ca.1943

That they were largely ignored in America is clear in many published histories, most of which Europeanized the rise and success of twentieth-century modernism.

The Griffins endured an inhibitingly conservative colonial Australia yet maintained a creditable architecture and city planning practice operated out of Castlecrag, a name they gave to their private suburb built above gentle shores beside an inlet of greater Sydney Harbor. It was the place where they settled after 1920 when he lost the director's post to supervise construction of Canberra. From 1914 into 1937 they designed buildings and landscapes, towns and suburbs, and university campuses that were located in Canada, the United States, Australia, and India. In 1935 Walter was invited to set up an architectural practice in Lucknow,

India. Marion joined him in 1936 and produced some brilliantly colored architectural drawings of their designs. But on Walter's death in 1937 she returned to Sydney, to Chicago in 1939. Then in the 1940s the Griffins involvement with knitlock reached an interesting conclusion.

In September 1945 the *Chicago Tribune* sponsored a "Chicagoland Prize Home Competition". Marion's entry made use knitlock blocks (made of "concrete or plastic") and re-employed the Pholiota square plan (but with a flat roof) as a housing "unit". Three units arranged corner to corner would comprise a house of 983 square feet, four units 1377, five 1622 (150.7 square metres). Corruption of the square plans to form larger houses was patent, the one room cottage was not expandable. (On return to Chicago Marion did not set up a formal architectural practice but assisted friends.) She entered a competition for a "Better Chicago" sponsored by the *Chicago Herald-American*, again in 1945, [7] She did not place in either competition.

During the 1920s the Griffins and Taylors were moving in opposed social circles; one in a Theosophy and Anthroposophy clique centered on Castlecrag, the other in conservative Anglican Sydney focused on commerce and colonial duties. In 1922 George Taylor was a delegate to the international "Peace City" conference in Brussels and a related League of Nations conference in Geneva. Sydneysiders Stowe and the Taylors traveled west via Perth, Western Australia, and the Suez Canal. (George failed to convince fellow delegates to disarm or to select Australia as the best site for a Peace City but he wrote a book about world peace.[8]) The Taylors' next trip to Europe was in 1924 and without a companion. George was an official delegate to the British Empire Exhibition while Florence gave a lantern slide talk in London about architecture. Enroute they spent ten days in the United States but little to nothing is known of those travels. Since they apparently did not visit Chicago or Los Angeles, the information would be irrelevant here.[9]

In December 1924 Walter and Marion began a sea voyage to America, their principal destination was Illinois homelands. Certainly to see Lloyd but also to satisfy professional curiosity, they visited Los Angeles twice in 1925, once before 7 January (perhaps via the docks at Long Beach enroute to Chicago), the second beginning around 22 February on their return route to Australia. Perhaps with their private hilly community of Castlecrag in mind they made a study visit to hilly Palos Verdes Estates suburb just southwest of Los Angeles. It had been designed in 1923 by Frederick Law Olmsted Jr. (the landscape) and Charles H. Cheney for subdivision layout.[10] Also, knowing of Wright's Los Angeles block buildings through American contacts and publications, they made a point to study the landscape of Olive Hill and at least two of the four textile block houses. Then in 1927, as previously mentioned herein, Walter commented on their construction similarities and differences to his knitlock system, but avoided aesthetic comparison. The article was illustrated with two of the Griffins' own photographs of the Freeman house.[11]

To be with his Austrian colleague, Schindler (twenty two years his junior), Neutra severed formal relations with Wright's Chicago and Spring Green offices and moved to Los Angeles early in February 1925. He had developed an appreciation of the variety of new American construction techniques and, after arrival, was thrilled to engage in the melèe of a greatening and opportunistic Los Angeles and to join its youth-driven society, enjoyed the comfortable weather, and gentle landscape environment. Prior to architecture studies Neutra had trained as a gardener in Vienna and went on after 1930 to become a landscape designer displaying considerable talent.[12] So he must have enjoyed Lloyd Wright's company. Through the deviant and debauched Loos, to whom he and Schindler were attracted intellectually if not socially (Schindler more so than Neutra), and perhaps through Otto Wagner (who Griffin met formally in Vienna April 1914), and before leaving Berlin for the U.S. in October 1923, Neutra came to know of the Griffins' Canberra design but not necessarily Walter's architecture.[13] We've noted that in ignorance Schindler had belittled Walter and disparaged his architecture in a March 1921 letter to Neutra, then still in Austria. Neutra had yet to move Erich Mendelsohn's Berlin office; that occurred late 1921 into 1923.

After winning the Canberra competition Walter's architectural work had been published in Switzerland and Germany, in Paris exhibited in the *Musèe des Arts Dècoratifs* and published in the *Gazette des Beaux-Arts* and in *L'Architecture*.[14] Neutra returned from military service in 1918 to work in Vienna. Then, while in Berlin, he would have continued to hear of Wright but probably not the Griffins. Yet after seeing Walter's work in the Chicago area, and then studying written and pictorial material about the knitlock system, the buildings and projects, and talking with the Griffins in February 1925, Neutra made his own judgement.

As to how Neutra learned of knitlock, this is the most likely scenario. In Wright's Spring Green or Los Angeles office people spoke of—or Neutra saw information about—knitlock as it had been presented by the Australian Ernest Stowe in 1919. Fascinated, Neutra was then able to talk with the Griffins (that February Wright Senior was in Wisconsin) and to obtain further details and photographs. That material came to Neutra's hand either on first meeting Walter or, more likely, after returning to Sydney Walter mailed it to Neutra. Griffin was, of course, also a landscape architect and had known Lloyd and his father for over twenty five years. Marion Griffin had known Lloyd since he was five years old.

The purpose of collaborating with the Griffins was to add material about knitlock to an article Neutra was preparing for a German book edited by architect and author Heinrich de Fries about—and with the full support of—Wright, and to a book Neutra was writing. His book discussed and illustrated a variety of new construction techniques in steel and concrete that he had discovered or had witnessed in Chicago and Southern California. As they were professionally interested in landscape and building architecture, Neutra and the Griffins were sympathetic colleagues. In any case Neutra believed European audiences were unaware of the American construction techniques and began writing about them just before settling in Los Angeles in 1925. The article appeared in de Fries's *Frank Lloyd Wright: Aus dem Lebenswerke eines Architecten* published by Ernst Pollak in 1926.[15] Neutra's book—or more exactly a pamphlet—with many illustrations, was published Julius Hoffman in Stuttgart in 1927 as *Wie baut Amerika?* (How does America build?).[16]

In that pamphlet Neutra wrote about and illustrated Griffin's knitlock, Schindler's structurally and aesthetically unique El Pueblo Ribera housing at La Jolla, Lloyd

Wright's slipform system for the Oasis Hotel in Palm Springs of 1923 to 1925, and Wright Senior's concrete block Storer house. Neutra was convinced by his own travels in the Southwest and by his hosts in Los Angeles that those buildings were inspired by the pueblos of New Mexico. To reinforce the argument of inspiration Neutra selected one of Schindler's exterior photographs of Taos Pueblo and another of some buildings at the once mysterious Siwa Oasis in Egypt (perhaps obtained from Mead or Gill whom Neutra had visited in 1924). The photographs were of traditional, indigenous yet universal adobe structures. We can recall that those regional types had been studied also by Gill and the well-traveled Mead. Neutra compared those plain unornamented mud buildings with Wright's textile block houses, his knobbly concrete adobes.

A very laudatory but confusing review of Neutra's book appeared in the New York *Architectural Record* written by europhile Henry-Russell Hitchcock, Jr. He described it as a "very important book" and compared the author with J.J.P. Oud in Amsterdam, Le Corbusier in Paris, and Mies and Gropius in Berlin. Neutra was "one of the less than half a dozen architects working in this country", Hitchcock asserted, "who are fully convinced . . . structure is the aesthetic crystallization of the engineering solution of the building problem". Hitchcock confusingly but interestingly mentioned that Neutra also discussed "the 'Knitlock' system of reinforced construction used in the last few years by Frank Lloyd Wright in his California houses".[17]

In reviewing the book for the Los Angeles *City Club Bulletin*, Pauline Gibling Schindler, Rudolph's American-born wife whom he had met in Chicago (possibly at Hull-House), believed it was "an interpretation of modernism and its expression in architecture . . . an affirmation and optimistic estimate of modern American civilization and architecture", and that Neutra had presented an exciting synthesis. The review also integrated some of Wright's theoretical utterances including bits of an essay of 1901 on "The art and craft of the machine". Pauline went on to say that a new style "is being created not mainly by the professional [American] architect, but by manufacturers and building materials and specialities . . . mass production by extensive machinery . . . ",[18] and by engineers, and so on. Realistically and as Hitchcock and Pauline should have recognized, on-site and labor intensive construction methods such as slipform concrete *in situ* or concrete blocks (or any other masonry) are not the technological stuff for the economical production of buildings, therefore not for inexpensive commercial buildings or workers' housing.

Conversely, architect and Historic American Buildings Survey researcher Robert C. Giebner described Neutra's Lovell House, also located in Los Feliz, as . . .

> *A prime example of residential architecture where technology creates the environment. The house is constructed of light steel framework, filled with standard window components. All parts of the structure were shop fabricated and transported to the steep hillside site—the structural skeleton being erected in just forty hours. The ribbons of wall are of thin concrete ["gunite"] sprayed against expanded metal backed by insulation panels acting as forms.*[19]

Fig. 8.5 Philip Lovell house in Los Angeles, Richard Neutra, Architect, 1927-29

In the experiential clarity of the design, in the making and in the appearance of the Lovell house, Fig. 8.5, Wright's textile block and tile buildings appear aesthetically regressive, technologically naïve.

But what of the working environment? In many ways the Los Angeles area's industrial and commercial development from about 1910 and through the 1930s was similar to the Chicago of his youth from the 1870s to 1914. For example, between 1920 and 1930, 1.5 million people settled in Southern California. The peak rate of 100,000 a year was achieved between 1920 and 1924. Another statistic: 6000 building permits were issued in 1918; in 1923 it was 62,548. Social historian Kevin Starr has put the extraordinary situation like this. By diverting mountain rivers,

> Water made imperial Los Angeles possible; but it was real-estate development and a phantasmagoria of attendant activities—buying, subdividing, building, selling, and finance—which within the decade of the 1920s propelled greater Los Angeles past the million mark, . . . the fifth largest city in the United States. An oil boom fueled this emergent economy, together with a tourist industry energized by Hollywood . . .[,] a City of Dreams, its boosters called it[.][20]

Avidly advertised throughout the nation, such a prospective city environment was in stark contrast to a troubled post-war Chicago with its soot-born air, surface grime, changing social order, horrid ghettos, and blatant commercial and political corruption.

In the many transitions from conducting an international war to promoting a civilian economy there was a certain amount of turmoil in America. But it remained strong, not suffering as had a torn Europe. Post-war consumerism soon increased and prosperity seemed certain. However, a dramatic increase in internal migration nationally, mainly by Negroes from former Confederate states to the northern industrial areas, lead to race riots. One of the most serious and deadly was in Chicago in 1919. Moreover, steel, mining and rail worker's strikes for better working conditions resulted in the rising importance—in the necessity—of labor unions. A related militancy by socialists and communists worried an increasingly conservative and intolerant nation. In words of the day, a "Red Scare" mentality set in. People with left-of-centre views were mistakenly aligned with angry anarchists. All were shunned, suspected, then set upon psychologically, socially, politically, even physically. We know something of the consequences Wright endured.[21]

Perhaps Wright was uneasy or anxious about the new Chicago and tired of old places and faces and battles. He was certainly eager to re-establish himself after neglecting an American architectural practice and suffering personal, social and professional debacles in Japan. He was desperate to create a viable professional practice in the not-so-angelic urban center on west coast America. After all, Wright might have thought, other architects from the northern mid-west area had recently emmigrated to Southern California and succeeded: Myron Hunt, Elmer Grey, Charles F. Whittlesey, James E. and David C. Allison, Canadian William J. Dodd, Peter B. Wight, and Gill among others. (James Allison, Whittlesey, Gill and Wright Senior had worked in the Adler and Sullivan office; Gill and Wright also in Joseph Silsbee's.) Fernand Parmentier was from Paris, France, via Chicago to Los Angeles … and so on. A rapidly expanding post-colonial Los Angeles was composed almost entirely of Euroamerican immigrants. Not a few were steadily and enthusiastically founding new traditions.

After many desperate years son Lloyd Wright was finally able to thrive as an architect in Los Angeles, but not father. Wright Senior had tried "very hard to get started on the West Coast. He had all kinds of projects going", Lloyd observed. "Considerable time" was spent by father in "trying to interest them [prospective clients] in his architectural schemes", Lloyd said. "[T]hey were very amused and entertained", but "unreceptive".[22] Personally promoting those schemes Wright called a "selling" campaign. The unreceptive included people in Madison, Lake Tahoe, Chicago, and Los Angeles where the failed campaign hurt most. The possibility of a partnership with his eldest son and namesake was lost.

One of those unsolicited campaigns was a scheme for an expensive resort offered to oil magnate Edward L. Doheny. In 1923 Doheny owned a large section of canyon lands in the Los Angeles hills. On speculation only, Wright offered Doheny some rather fussy, elaborate, labored, if visionary images of a future but rather ill-defined resort. It would include up-market housing on earth ledges cut into hills or spread across shallow depressions. Of those designs the best known are a set of colored perspective drawings, the largest vaguely reminiscent of a traditional Chinese landscape with many detailed views of atypical buildings held visually by mist and trees and land forms. A bevy of rather imprecise design drawings that together formed a panoptic view of a few arched bridges and small to large buildings. The walls of some buildings looked as if Lloyd's slip form concrete or father's textile tiles were to be used. All were unusual, aesthetically and structurally, and regressive.[23] Among them were three fancy houses as specimens for an expensive market.

However, dramatic national events played against Wright's proposed resort. By 1923 the sordid activities of President Warren G. Harding's administration could no longer be concealed. The more newsworthy scandal was a shady deal by Secretary of the Interior Albert Fall who accepted a bribe of $400,000 to issue secret leases for access to oil reserves at Elk Hills, California and Tea Pot Dome in Wyoming. The leases had been given to oil tycoons Harry F. Sinclair and Doheny. Fall was convicted but oddly, the two oil men were not. Depressed, Harding died of a heart attack in 1923. Doheny was in no mood to consider a posh resort let alone contemplate Wright's entreaties. However, it was soon thereafter that a chastened Doheny began a program of philanthropy for Southern Californians.

We should recall another setback. A series of bold designs, some of which could have employed textile blocks, were prepared by Wright for his own speculative venture to build a holiday resort on the shores of Emerald Bay on the California side of Lake Tahoe. The tiny bay was first settled by Euroamericans in 1863 with a summer home for Ben Holladay. A rough small resort was built in 1884. In 1892 the William Henry Armstrong family purchased a major portion of the bay's land, about two hundred acres that included Fanette Island, the only island in the lake. With a water fall, magnificent pines and cedars set off by shear granite cliffs the bay was and is a scenic wonder. Part way into negotiations Wright was encouraged by a letter in which Jessie Armstrong tentatively offered "cooperation", as the architect put it.[24] Apparently Wright also hoped to entice others to join in financing the summer colony promoted and designed from mid-1923 into 1924. His designs included cruising houseboats, barges, and a bridging walkway on pontoons to Fanette Island. Fortunately for the future preservation of the bay and is surrounds, to which the Armstrongs were committed, the project faded. Jessie Armstrong was candid: "he [Wright] had peculiar ideas of architecture; everyone didn't respond to them".[25]

Nearly all of Wright's proposals or commissions in Japan had not been realized or did not receive his personal attention during his residence 1917 to 1922. Yet, as he had done in Japan, from 1922 into 1924 Wright tried to interest Los Angeles businessmen in proposals for a new "merchandise" store, a motion picture theater, an office building, then a "glass, copper and concrete" skyscraper; each for unspecified sites downtown. As in Tokyo, people in Los Angeles smiled politely, Lloyd recalled, but said no.

One of few architectural commissions was obtained in 1923 from prompters of a new clubhouse for the Nakoma Country Club. As mentioned, it was a typical speculative real estate project located just west of Madison. Again the project failed to mature but the club prospered. Another commission was another house for an uncertain and fickle Aline Barnsdall, this for a steep site in a newly set up and up-market Beverly Hills suburb. That 1923 project, at one time to have employed concrete blocks and textile tiles, was active February through May, inactive by July. A large "compound" and house was commissioned in 1923 by Chicagoan A.M. Johnson for a site at Grapevine Canyon, in the Mojave Desert north of Death

Valley. Things progressed to a stage where a good set of preliminary drawings were prepared for a extensive house. Apparently, at one stage the buildings were designed to employ concrete blocks, but in 1924 Johnson pulled out.

At the time of their association Johnson was director of the National Life Insurance Company. He hired Wright in 1924 to design a new company headquarters to be sited on one-half of a city block in North Chicago. A massive office building devoted in the main to rental space came to life on Wright's drawing board. It was to be a concrete structure of cantilevered floor slabs, a structural technique initially designed for the six story section of the Tokyo hotel and for the skyscraper of "glass copper and concrete" project of 1923. A prefabricated metal system designed for the exterior skin and interior partitions was similar in detail to that prepared for the 1923 skyscraper. Johnson abandoned the project sometime in 1925.

In 1923 Darwin D. Martin asked Wright to design a house for an unknown site in Buffalo, New York, for Dorothy Martin Foster, husband of James and daughter of Darwin who in 1903 had commission the Larkin building. On receipt of the preliminary drawings, see Fig. 8.6, that showed an oddly, highly contrived facade of clay brick masonry and/or precast concrete block masonry, not textiles, Darwin withdrew.[26] We can usefully compare it to Neutra's Lovell house and the houses of Schindler 1921-1925.

Martin Saches, who taught at Deep Springs College just north of Death Valley in California asked Wright to design a small house for his bride. It was not a formal commission and apparently Saches never received drawings. However, Wright did sketch a floor plan and a nice color perspective in 1924 that displayed a delightful two bedroom house, adobe in character. It might have been designed for ordinary concrete blocks or for textile adobe. Also, in the perspective there appeared to be colored tiles of some sort applied to the upper reaches of the principle exterior walls.[27]

When one surveys these many designs of the period 1922-1924, other than the four upon which we have concentrated, it would be fair to say that Wright was trying, rather desperately, to attract customers by creating the startlingly new or the daringly different, and thereby prove not just the attractiveness of artistic and social eccentricity but of his genius. With the exception of Johnson and Ennis, the clients he obtained, however, were not holders of conservative big money but those on the political left margins of middle income society, those politically unacceptable, those often avoided by an impolite community stuck in convention and fearing the unknown.

As early as 1921, Wright experienced the early stages of the disillusionment that Lloyd was to observe. "Los Angeles was a strange new field for me", Wright Senior said in a letter from Spring Green to Aline Barnsdall. "Apparently they did not want me out there and do not want me now". He knew people "there" thought of him as a "freak architect and his fool-work", and that they spread "stories about his private life on the Q.T.", to use his words. We can remember what he despairingly wrote the following year to "dear Dr. Berlage" in Amsterdam. In that letter he went on to say that clients had to be of "imposing independent thought and action" and expect to suffer "some pains" for their association with him.[28]

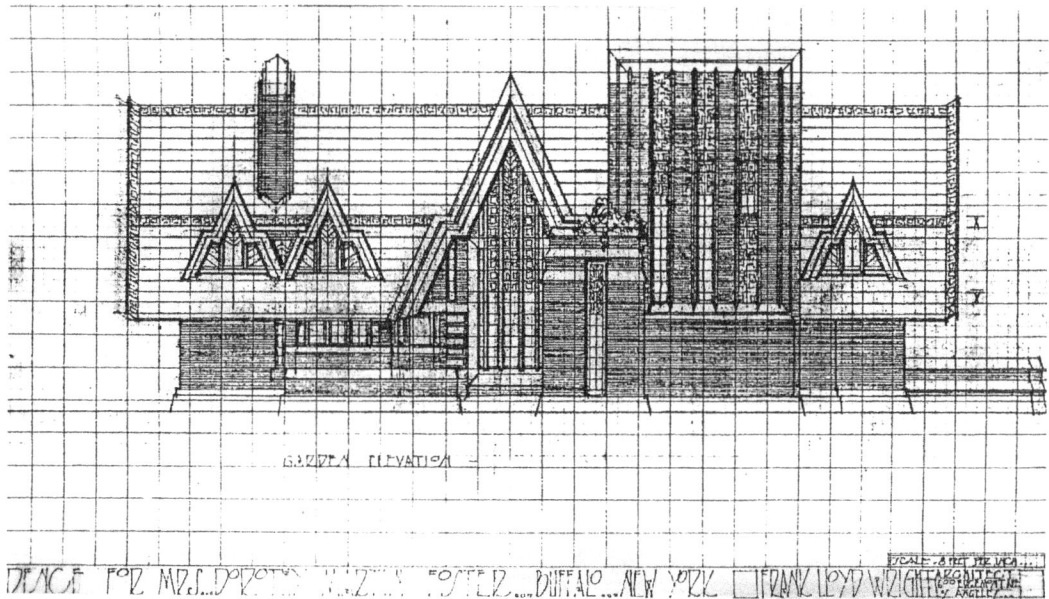

Fig. 8.6 Foster house for a site in Buffalo, New York, a project of 1923

Disenchanted, discouraged and beaten, he closed the Los Angeles office on 23 September 1923 and shifted up north to the Spring Green house, Taliesin. Finally he gave up the Southern California adventure and returned permanently in 1924 to the familiar cool grey-green rolling hills of rural Wisconsin.[29] Only occasionally thereafter did he return to Los Angeles.[30] Historians Burchard and Bush-Brown offered a sympathetically balanced, almost tender observation of this, another self-imposed but nonetheless tragic moment in Wright's life. Being largely ignored, the two historians said,

> he [Wright] turned inward at Taliesin East, [Spring Green,] Wisconsin, creating there a beautiful house and farm, where he surrounded himself, feudally, with young disciples, mostly from abroad where his reputation never faltered. It was a time for reflection, partly self-enforced by rebellion against society, prolonged by the rampant social urge toward standardization and advertising and extended still longer by the depression.[31]

A short yet wonderfully creative period had closed.

Wright applied the new textile concrete adobe to a variety of buildings and unrealized projects for the Southwest. Once back on family soil and reacting to resistant prejudices, he designed concrete adobe buildings for locations all about the American landscape, each site with different regional, climatic, and social conditions. Concrete adobe buildings built or proposed were planned for sites in Tulsa, Oklahoma; Chandler and Phoenix, Arizona; Madison, Wisconsin; Death Valley; Buffalo, New York; and Denver, Colorado. Those designs reveal that by 1924 he had indeed discarded the theoretical proposition of regionalism. Perhaps he realized that, ignoring the technique of textile blocks, the Freeman and Storer houses had

Fig. 8.7 Larkin Company Administration Building in Buffalo, New York, 1903-05

otherwise transcended regionality. Being conceptually more romantic than not, they were not abstractions but nonetheless pregnant with potential theoretical extension, more responsively organic.

In the years between 1898 and to around 1917, and as a general observation, Wright's ornament was abstracted from prairie flora, sumac a favorite, and most obviously seen in designs for leaded colored glass windows. But after returning from Europe and from trips to the southwest and talks with Gill and Lloyd, Wright's ornament came out of many exotic sources and assumed a universality. But again change. In the 1920s occasionally, but around 1930 surely, all references to regionalism and exoticism were ignored.[32] Yet consider this. When we harken to Wright's nondomestic buildings of the period 1898 to around 1914, the paradigms are the Larkin building, Fig. 8.7, and Unity Temple. Here we note the utter clarity of parallelepipeds, of nonreferential rectilinear forms carefully massed as they exactly articulated functions. There are only hints of regional flora on sculpted column capitals. Even colored glass leading was rectilinear in contrast to Sullivan's flamboyance and Europe's curly Art Nouveau.

With Olgivanna, a young, beautiful, intense, and rather mystical wife (thirty years his junior), whom he had met in 1924 and (after receiving a divorce from Miriam in 1927) married in 1928, a newly found optimism emerged. Jointly they directed all energies not only to reviving an ailing architectural practice, but to warning the nation of a wicked political and artistic internationalism emanating from Europe, one not of, by and for America. He began a verbally combative campaign centered on arguing the correctness of principles that he believed identified America, and other principles and personal actions he believed would improve her condition, especially in the throes of The Depression. His verbalized thoughts wandered with catholicity and imprecision. In many ways it was a case of "Wright versus America". Also begun immediately after marriage were activities to recycle Wrightian mythologies and a biography as a first step toward refurbishing a career. Practicalities focused on creating a new legend based on reviving for public consumption the history of over forty years of extraordinary artistic activity. Further, husband and wife prepared an educational program for their own design school where the Master presided. He offered to any America who might listen, ideas not only verbally but in the form of architectural or city planning designs for the present and future. All this, he said, to attract clients: by 1938 they had succeeded.[33]

CONCLUSIONS

> **I am struck by the fact that he [Wright] was careful to explain in his Wasmuth commentary [of 1910] that the houses therein were borne of the prairies. The inference being that one could not fully appreciate the architecture without an understanding of its generative landscape context. [However] So far as Chicago was concerned, to be "regional" was to be "national" or American in as much as many believed Chicago (vs "olde-world" Boston or New York) to be the most emblematically-American city of all.[34]**
>
> — **Christopher Vernon**

Historian Henry Schmidt provided another neat summary that here serves as a necessary clarification of a parallel phenomenon. In the literature and experiences of the period discussed herein, one dominant theme, Schmidt said,

> *was the search for a New World community which entailed a flight from the machine age and an encounter with nature and simplicity—a variation of the experience [of] "pastoralism" in classic American literature. In different degrees, these works manifested the interest in socialism, the rebellion against Philistine America, a New World aesthetic, and the return to native origins that intersected the American intellectuals of the period....A pervasive image... is that of escape from the monstrous sophistication of America to an Edenic primitivism where human faculties might be purified and begin to function naturally... [a] flight from "civilization to clamorous and exigent...".*
>
> *In an abstract sense, to go from America to Mexico was to leave the realm of the machine and enter the realm of art.[35]*

Wright was emotionally, socially, intellectually, and professionally attracted to those persuasions. But we now understand that he refused their direct infusion into life and art. Rather, he manipulated them in highly idiosyncratic manners to meet his advantage.

A few months after the appearance of Tselos's article of 1953 about exotic and Mayan elements, Wright acknowledged there are many influences on the "mind of youth". They "may" be "sources of inspiration", but also they "might go backward as well as forward", back to the source of the source, back to the primordial. He frankly described "the article by Professor Tselos" on decorative eclecticism (borrowings) as yet another "haphazard guess". It was like "gazing into the crystal ball to see whence came the image:" the *image*, please note. Not the Idea. Not the Concept. Not the Intention. He added that many people "seem to be cutting out the head of the drum to find whence comes the sound. . . ". In elliptical prose Wright went on to acknowledge that Mayan architecture was "based upon glorified abnegations of man to authority". Nevertheless it was "intrinsic to Time, Place and Man". He was satisfied that over succeeding centuries the Central Americas had under gone changes such that in 1953, "Now—Freedom is from within;"[36] now it arises from— and resides in—the precious Individual. (It was more properly an expression of hope than an observation of reality.)

We can recall that while ignoring their sophistication Wright had said that pre-Columbian buildings were but one "primitive basis of world-architecture". But always, "primitive" means prime, i.e. first. Douglas Haskell had put it nicely. Anyway, Wright continued his 1953 ruminations:

> Only with that understanding could I have shaped my buildings as they are. Yet, of all ancient buildings, wherever they may stand or whatever their time, is there one of them suitable to stand here and now in the midst of our time, or America, our machine-age technique? Not one.[37]

The use of the term "ancient buildings" is more satisfactory than "primitive". And so with these few candid words Wright once again rejected borrowing or redesigning the exotic, sophisticated buildings—those images—of the Maya and other people. More to the point, he was urging readers to consider a more philosophical position, one that would strengthen architectural theory. The source and inspiration of all architecture is its primal foundations, its "primitive basis", its nature-dependent and social determinants. Private cultural images (e.g. shape, color, decoration) are elaborations, perhaps additions (dehumanizing encumbrances Jean-Jacque Rousseau had said), yet fundamental to social identity, pride, clan, purpose, even stability. When writing theoretical stuff for a lay audience while resting under canvass lit by sunlight in quarters at his Ocotillo "camp" site near Chandler, Arizona, in 1929 Wright used reflection in example: "On Midwestern prairies I build . . . houses that proclaim the prairie's quiet level,—the third dimension evidence as unbroken roof-planes likewise lying in similar repose,—as human *shelter* [he emphasized] . . . ."[38]

Excluded were nondomestic buildings so patently not in repose. Houses referred to were those such as Warren Hickox of 1900 and for the Millards of 1906. They were the buildings that in 1924 an observant J.J.P. Oud correctly described as the foundation of the "Wright School:" Oud's words.[39]

In 1957, in his last book, Wright wrote that "No school exists without something to teach". He went on to say that he "was not aware that anything like a 'school' had existed", and then went on, as only he could, to prove it had. "A small clique soon formed about me, myself naturally enough the leader", he said in conceit. It was an insult to the memory of Louis Sullivan. Regardless he listed his "followers": Cecil Corwin, Robert Spencer, George Dean, Hugh Garden, Myron Hunt, Dwight Perkins, Richard Schmidt, Howard Shaw (strangely), Gamble Rogers, Birch Long, Max Dunning, "an ebullient Italian, [Adamo] Boari by name", the Griffins, William Drummond, George Willis, Andrew Willatzen, Frank Barry Byrne, Albert MacArthur, a reluctant George Grant Elmslie, and "others came and went as time went along". Will Purcell was certain that in 1911 there were "young men who were following Wright's footsteps".[40]

Not all followed one of the Wright Styles. Some, he moaned, remained tied— or switched—to historical revivalism like Charles E. White and Byrne. Nonetheless most "fell in with the idea", Wright said, and then boasted that "I [he] became original advisory exemplar to the group".[41] They also included Thomas E. Tallmadge, Vernon S. Watson and Harry F. Robinson. Most of these architects drew on Wrights horizontal domestic designs while others were happy to emulate his non-domestic buildings; some both styles.

Prairies throughout the world can inspire thoughts about their region but only a person "with something to teach" can conduct a school of students and followers, and then only by creating both a rational premise and architectural products that define that premise. There was no prairie school just as there was no desert school of the southwest. There was a Wright School whose teachings and products explored designs for *both* domestic (regional and nonregional) and nondomestic (nonregional) architecture.

Wright believed his buildings for the American Southwest, as outlined earlier herein, were not an alteration of his aesthetic mission.

> *The effort in California and Arizona? [Wright asked] ... these desert buildings too are naturally elemental in form. ...*
>
> *Unhappily, my critics ... must continue to see Egyptian, Mayan, Chinese, Japanese, Persian, Moorish. Not one motif of the sort can they fairly fix in these buildings for such were never in my mind.*
>
> *Did I prefer them lean,—sun-defiant,—ascetic? They might be so,* honestly *[he asserted] ....*[42]

Wright was not the only architect to safely mine but not mimic historical precedents in a search for inspiration that would responsively lead to modern architecture. For examples, his contemporaries Bernard Maybeck in San Francisco and the Greene brothers in Los Angeles have been studied by historians who "decoded" (as they like to put it) the work of these three architects. They found "a diverse melding of Queen Anne, Stick and Shingle, Arts and Crafts, Swiss, Bavarian, Scandinavian, Tibetan, Bhutanese, Sikkinese, and Japanese influences";[43] meld as to blend. But our encoded eyes can deceive.

It is clear that Gill's architecture took something from southwestern Native Americans, Native Northwest Africans, and from other similar architectures; and he accepted the white round arch as more universal than merely a symbol of Spanish colonial: that was the sum of his useable investigations. They evolved and matured to become a potent reductive theory, a gift of Gill's imagination.

As abundantly and openly revealed in Schindler's built or proposed designs it is safe to say that he had rejected Wright's aesthetic responses. He found them regressive yet accepted Wright's words, his theoretical utterances at face value. He (and Neutra), as had other central European architects, could see that Wright's built product did not match his published theories. Neutra proved his talent with the Lovell house and as much as he benefitted socially and professionally, he was now no longer dependent on Wright or on the Schindlers and their politically fevered arts entourage.

AND THE TEXTILE BLOCK HOUSES?

What of the exterior and interior surfaces of Wright's textile block houses? Sullivan and Wright seemed to be in theoretical accord with Owen Jones aphorism that "Construction should be decorated. Decoration should never be purposely constructed". Writing in the late 1920s Wright rather confusingly elaborated with emphasis:

> *integral ornament [is] the developed sense or the pattern-of-structure itself. Integral pattern is structure-pattern made visibly articulate....*
>
> *[The] expression of the structure as pattern and the nature of materials... may be taken further than bare need..., [taken] into the higher realms of Imagination.*[44]

The pattern-of-structure is evident in the host of Wright's buildings prior to 1924, some exaggeratedly "articulated". We can identify examples such as houses for D.D. Martin, Shaw (Montreal, Canada, 1906), Storer and Ennis, the church Unity Temple, the office building for the Larkin Company, the prescient Midway beer garden, even the regressively effusive Tokyo hotel.

While functional articulation is plainly evident in the floor plan and exterior massing of Barnsdall's Hollyhock house, the structural pattern is not. Moreover, the only exterior decorations are two types of precast concrete elements in relief sitting on the lintel-high string course, and they are "purposely constructed" as an additive. The detail and character of the interiors and their architectural decoration are typical of Wright's houses in the upper mid-west before 1914 and in this case inadequately respond to the exterior. It is a hermaphroditic building.

On the other hand, the interior and exterior bearing walls and square posts, boldly exposed or expressed, of the textured concrete adobes for Millard, Freeman, Storer and Ennis each have an obvious ornamental integration. Such an integration and its visual character are found in certain traditional Southwest Asian architecture; another of those exotic sources. Shortly after the four Los

Angeles houses were built Wright acknowledged that source when he said that the four houses

> *stand delicately perforated like a Persian faience screen*
> *or lie low and heavily in mass upon the ground.*[45]

Elsewhere he reiterated the evidence of that characteristic, saying the Los Angeles houses with their "jewel-cut surfaces" would permit "the fancy of the Persian or the Moor".[46] A Moor was then understood to be a person resident in Northwest Africa, like Morocco, where traditional buildings sit heavily "upon the ground". So too those in Persia and at Taos and at Walpi.

It is also clear that the immediate and profound influence upon Wright, i.e. just before designing the textile block houses, was Gill's severe planer and volumetric reductions post-1906. In nearly all cases they were equally articulated. They had been derived conceptually from a theoretical attitude that Wright thoroughly appreciated and, in a rare display, publicly praised if rather indirectly in words, more profoundly in architecture.

With the textile block houses, however, Wright believed the structural pattern jubilantly reflected "the higher realms of Imagination". This was achieved not only by a clarity of structure in floor plan and three dimensional form, as Gill had done so obviously, but by pressing ornamental designs into the surface of wet concrete, then to be not sun dried but air dried. In the case of the Ennis house they collectively and daringly expressed the articulated functions.

(None of these critical musings can be applied to the theoretical impulse for Griffins' knitlock and Lloyd's knit-block, or to the purpose and character of their buildings.)

While the Arizona Biltmore Hotel of 1928 concludes the concrete adobe experiment. Architectonically it was regressive in all respects. More properly the Ennis house of 1924 became the ultimate expression of the experiment and a furtherance of articulation. While it identifies a conclusion, the work of Schindler from 1922 onward and Neutra 1927 on, exemplify the resurgence of a responsive holist architecture, an architecture of structural clarity and revolutionary aesthetics first practiced by Gill. Wright reacted four years after the Lovell house was complete, but that story is told elsewhere.[47]

+ + +

The non-chronological and therefore non-linear—and to a lay audience the irrationality—of creative activity is perplexing, of design even more so. Yet, so much of published art and architectural historical research is devoted to the assumption of a progressive linearity founded on European precedents. We've rigorously disclosed that too often they are exposed by sets of what Anthony Alofsin neatly described as "superficial visual analogies". Since architecture is a joint enterprise of client, architect, builder, and user, an aesthetic clarification of the conceptual intention is paramount to cognition and appreciation. Prompted intelligently by

Gill, supported by Schindler and Lloyd, Wright *intended* the Los Angeles houses 1919-1924 to clearly state they were aesthetically antithetical to antique and historical Mediterranean precedents and those of modern Europe. Yet, he wanted them to be a vigorous continuation *and* extension of primary fundamental determinants, and of his own precedents as they might aesthetically define ideas. He would have liked his audience to perceive such a continuity and originality in spite of apparent visual disparities. He intended, therefore, that the houses be modernized, Americanized, and Wrightian realities of the character of American Southwest indigenous architecture. The earliest house was obviously romantically tedious, nearly imitative. The kindergarten was an ill-conceived aberration. The other four houses, however, announced that noble intention. They did so in a manner that proclaimed a timeless universality (even with "jewel-cut surfaces") that he took from Gill. All this was then and is now experientially apparent.

A buoyant and introspective Wright wrote to an admiring Lewis Mumford in 1929 about many things, including the prejudice of precedent:

> *The New in art [Wright said,] is always formed out of the old....*
>
> *Now as a student of the mystery of form, I have found an outlet for all my energy....*
>
> *Nothing confounds the critic, it seems, like common sense or straightforward dealing of any kind. He is taken in by surface indications ... and fails to see it [architecture] as a reality beneath an exuberance.*[48]

Indeed. The earth and the past, for the present is past, is all we have at hand to form the future. The Hopi are correct. And indeed, to understand a building we must go beneath "surface indications" and beyond aesthetical words to find the architectural reality.

Having not experienced or being unfamiliar with or not understanding the architectural design process *and* the actual construction process, practicing too narrowly an art historical methodology too reliant on surface ornament and mere visual comparison, failing to objectively weigh the evidence or read texts or engage in an excessive devotion to analogies, cause historians and observers—especially the verbalists—of architecture to perpetuate fictions that distort revelations of past events, avoid purpose, constrict biographical and art narratives, prejudice evaluation, and consequently diminish the potential of historical lessons. Equally obvious, the designer is treated unfairly by verbal vagaries and half lies/half truths.

Unsubstantiated assertions—and an uncritical and cloying adulation!—have no place in biographical and architectural studies of any architect, but especially of Wright, the most public yet enigmatic of great architects. They diminish all that is worthwhile in our expectations of the discipline of architecture. The historian's task is—my task has been—to "question, criticize, explore and correct the record", as Thomas Hines has said, "without losing the poetry and the richness of Wright's achievement".[49]

NOTES

1. Epilogue: Cormac McCarthy, *Cities of the Plain* (New York: Vintage, 1998), p. 286.

   Schindler's drawings for Kings Road House are reproduced in Gebhard (1993), vol.2; walls are detailed along with photographs of them in Ford (1990), pp. 290-294; see also Patricia Kucker, "Framework: construction and space in architecture of FLW and Rudolf Schindler", *Journal of Architecture*, 7(Summer 2002), pp. 180-190; Kathryn Smith, *Schindler House* (New York: Henry N. Abrams, 2001), reprint Los Angeles: Hennessey & Ingalls, 2010; Robert Sweeney,"Life at Kings Road as it was 1920-1940", in Smith/Darling (2001), pp. 87-115; Elizaabeth A.T. Smith, "R.M. Schindler: an architcture of invention", in Smith/Darling (20001), p. 25f.

   For a sample of Schindler's architecture see Jin-Ho Park and Lionel March, "The Shampay Houses of 1919, Authorship and Ownership", *JSAH*, 61(December 2002), pp. 470-479; Jin-Ho Park,"Schindler, Symmetry and the Free Public Library, 1920", *Architectural Research Quarterly*, 2(Winter 1996), pp. 72-83.

   For social and artistic insight see John Crosse, "Pauline Gibling Schindler: Vagabond Agent for Modernism, 1927-1936", <socialarchhistory.blogspot.com/2010/07/pauline-gibling-schindler-vagabond.html>, accessed November 2010.

2. Kings Road house is now owned by Friends of the Schindler House with support of Austrian Museum of Applied Arts, or MAK, and is open to the public.

3. James Steel, *How House: Architect Rudolf Shindler* (New York: John Wiley 1996). The How house, located in the Silver Lake district of Los Angeles, used moveable wood formwork with scribed joints between concrete pours on two exterior walls and some concrete masonry. The concrete rises out of the earth to embrace horizontal redwood siding with relief joints. Roof structures are exposed on the interior. It has been nicely renovated; the initial landscape by Neutra.

4. Schindler to Lloyd, 9 April 1923. as quoted in Kurt G.F. Helfrich, "Contextualizing 'space architecture' what the Schindler Archive reveals", in Smith/Darling (2001), p. 151; Ribera's structure is detailed in Ford (1990), p. 296.

5. As quoted in by Kurt Helfrich, in Smith/Darling (2001), p. 156.

6. Gebhard (1993), vol. 4, plate 3068.

7. Pamela Hill, "Marion Mahony Griffin: The Chicago Years", in Charles Waldheim and Katerina Rüedi Ray, editors, *Chicago Architecture: Histories, Revisions, Alternatives* (Chicago University Press, 2005), pp. 52-161. Andy Rebori was a juror. The sheet of drawings were altered and photos of a couple of knitlock blocks added when made part of a manuscript for Marion's unpublished "Magic of America", Griffin (1949).

8. George A. Taylor, *A world of peace: its advantages and how to win them: a study of nationalities and the Third Assembly of the League of Nations, at Geneva, 1922 . . . with portrait sketches by the writer* (Sydney: Building, 1922), 3rd edition or printing 1923.

9. Taylor (1924); Michael Roe in ADB, vol. 10; Freestone/Hanna (2008), pp. 52-57.

10. Fogelson (1967), pp. 157-161.

11. Griffin (1927). While employed in Wright's office the Griffins probably knew the Millards personally when working on their first house in Highland Park, Illinois (1905-1906).

12 Kenneth Frampton, "In Search of the Modern Landscape", in *Denatured Visions: Landscape and Culture in the Twentieth Century* (New York: Museum of Modern Art, 1991), pp. 48-51.

13 Christopher Vernon generously shared his research into events around 1925. See also Vernon, book review of De Long (1996), *JSAH*, 57(September, 1998), pp. 353-355.

14 Christopher Vernon, "'The silence of the mountains and the music of the sea': the landscape artistry of Marion Mahony Griffin", in Debora Wood, editor, *Marion Mahony Griffin: Drawing the Form of Nature* (Evanston: Mary and Leigh Block Museum and Northwestern University Press, 2005), p. 14; and Vernon (1996), pp. 147-148.

15 De Fries (1926). For an evaluation of de Fries (1926) see Anthony Alofsin, "FLW and modernism", Riley (1994), p. 56,n45, 49.

16 Richard Neutra, "Eine Bauweise in bewerhtem Beton an Neubauten von FLW" (A reinforced concrete construction method in Wright's new buildings), in De Fries (1926), and reprinted in part in Neutra's *Wie baut Amerika?* See also Richard Neutra and El Lissitzky, *America* (Vienna: Anton Schroll, 1930).

17 Henry Russell Hitchcock, Jr., review of *Wie Baut Amerika?*, A Record, 63(June 1928), pp. 594-595.

18 Schindler as quoted in Hines (1982), p. 65.

19 Peter Lizon, "Lovell Health House", sheet 1 of National Park survey no. CA1936, *Encyclopedia* vol. 2, p. 797. The New York City boy Morris Saperstein changed his name to Philp Lovell when he migrated to Los Angeles after receiving a chiropractic degree in Kirksville, Misssori. Late 1920s and through the 1930s he wrote "Care of the Body" for the *Los Angeles Times*. He and wife Leah Press Lovell wrote a series of books and pamphlets on health issues and alternative medical cures.

20 Starr (1990), pp. 68-69, cf. Chapter 8. See also Greg Hise, *Magnetic Los Angeles Planning the Twentieth-Century Metropolis* (Baltimore: Johns Hopkins University Press, 1997); Fogelson (1967); De Long (1996), pp. 16-19.

21 Johnson (2005), chapters 13-15.

22 L. Wright interviewed by Gebhard/von Breton (1971), pp. 27-28. It was in the 1920s that, while America shunned him, Europe once again took an extraordinary interest in Wright's architecture, see Langmead/Johnson (2000), chapters 4-7; Alofsin (1999).

23 For text and illustrations see De Long (1996), pp. 21-29; Riley (1994), p. 212. Doheny eventually built a massive Gothic Revival mansion, completed 1928, to a design by architect Gordon Kaufman.

24 Wright (in Wisconsin) to L. Wright, 6 May 1924, some drawings were enclosed and the letter also noted receipt of a "letter from Jessie Armstrong yesterday suggesting cooperation", copy LWPapers. That part of the bay is now a state park named after its donor, lumberman Harvey West.

25 De Long (1996), p. 64.

26 De Long (1996), p. 65. The Armstrongs sold the land in 1928. For photographs, map drawings and general history of the bay see De Long (1996), pp. 47-67; Levine (1996), pp. 169-173; on Nekoma see Sprague (1990), p. 208, chapters 9 and 10; on Johnson see Levine (1996), pp. 173-185; De Long (1996), pp. 66-79; on Foster see Fisher Fine Arts, *FLW*, (London et al, 1985), catalog items 33-38.

27 Pfeiffer (17-32), plates 80-82; Levine (1996), p. 459,n52; De Long (1996), pp. 70-71.

28 FLW to Barnsdall, 27 June 1921; and FLW to Berlage, 30 November 1922, in Pfeiffer (1984), p. 34, 54.

29 Kamecki Tsuchiura and his wife worked for Wright who had invited them in March 1923 to join him and William Smith in Los Angeles. In December 1923 the Tsuchiuras and Smith were sent back to Spring Green leaving only Lloyd Wright in Los Angeles, not as an employee but as an occasional contact. During 1924 and 1925 the following were also working at Spring Green: Werner Moser and Neutra both from Austria, and A. Feller from Germany; no Americans. The nature of their employment is unknown but are indicative of European interest in Wright. The Tsuchiuras return to Tokyo in November 1925 parallels a waning interest in Japan. See "Kamecki Tsuchiura, Draftsman from Japan", in Edgar Tafel, ed., *About Wright* (New York: John Wiley, 1993), pp. 93-95.

30 Johnson (2005), passim.

31 Burchard/Bush-Brown (1961), p. 369.

32 From thoughts shared with landscape architect Christopher Vernon.

33 Wright's career in the 1930s is the subject of Johnson (1990); and see an exuberant Toker (2003).

34 Vernon to this author, 9 May 2007.

35 Schmidt (1978), p. 345; also cf. Oles (1993).

36 FLW to Grant Manson, 18 June 1953, as quoted in Pfeiffer (1984), p. 115. Cf. Grant Hildebrand, *The Wright Space* (Seattle: University of Washington Press, 1991), chapter 5.

37 FLW to Robert Goldwater, appendix to Dimitri Tselos, "FLW and World Architecture", *JSAH*, 28(March 1969), p. 72.

38 Wright (1929a), pp. 92-93.

39 J.J.P. Oud, "The Influence of FLW on Europe", Wijdeveld (1925), p. 86.

40 As quoted in Vernon (1996), p. 139.

41 Wright (1957), pp. 34-36; those paragraphs were an elaboration of a discussion about colleagues when he began practice in Chicago as contained in the first edition of his autobiography Wright (1932) pp. 128-129, only slightly altered in Wright (1943), pp. 130-131. Perhaps Robert Bruegemann should have included the Prairie School myth when he wrote a short essay about the "Myth of the Chicago School" in Waldheim/Ray (2003), pp. 15-29.

42 Wright (1929a), p. 93.

43 As recounted by Starr (1990), p. 188.

44 Wright (1932), pp. 358-359.

45 Wright (1928), p. 102. "Faience" is glazed tinware or earthenware or porcelain. Wright supplied some photographs and text to Onderdonk (1928).

46 Haskell (1928), p. iiv.

47 Cf. Johnson (1990); Toker (2003).

48 FLW to L. Mumford, 7 January 1929, in Pfeiffer (1984), p. 143.

49 Hines (1995), p. 475.

Appendix One

IRVING GILL ON ARCHITECTURE

Introduction: Gill spoke only once for publication about his theory of architecture and that was in a 1916 article only recently reprinted.[1] There is no need to reprint the reprint here. The following three pieces, however, are from published essays by other authors that interpreted Gill's ideas as they were revealed to them on site visits or during interviews with the architect conducted before 1916. The first two were first presented in the New York magazine *The Craftsman* accompanied by illustrations. The second also illustrated was in the Minneapolis *The Western Architect*. Each has not been published since. During 1905 Eliose Roorbach attended the Mark Hopkins Art Institute in San Francisco

+ + +

"OUTDOOR" LIFE IN CALIFORNIA HOUSES, AS EXPRESSED IN THE NEW ARCHITECTURE OF IRVING J. GILL
BY ELOISE ROORBACH[2]

[The] deliberate simplicity [of Gill's architecture] cannot possibly be overlooked. It compels attention. It calls to mind Schiller's observation that "The artist may be known rather by what he omits."[3] The architect, Irving J. Gill, with pioneer courage resolved to go back to certain fixed principles like the line, square and cube, and to build from them with as little deviation as possible, omitting everything useless from a structural point of view. He came to see great beauty in straight lines. He grew to love them, to combine and recombine them, and to merge them. He studied the charm that lies in perspective and applied it to his lines of roof, walk and wall. He saw that ornament was a non-essential. So he determined to make his houses depend for their beauty entirely upon the relation of line to line, of surface to surface, proportion to proportion, and then plant vines and flowers to furnish decoration.

+++

RADICAL WESTERN ARCHITECTURE[4]

Whoever speaks of Irving J. Gill's work . . . must dwell upon his radical views of design as well as of construction.[5] Architects more than any other creative workers, perhaps, are forced to work somewhat within traditional limits. Their designs must be governed by consideration of practical things, such as cost and endurance of material, size of lot and the personal wishes of the owner. Their work must be beautiful and original in design, permanent in construction, able to withstand fire, time and the elements. Without the tremendous cohesive force of imagination, these herculean tasks could not be performed.

Desiring to do his part toward creating a new domestic architecture, Irving Gill threw aside all the conventional standards of known styles and began with the three first principles, The square, the circle and the line. Working with these fundamental powers and ridding his mind of all accepted standards of ornamentation used to cover up defective lines, he saw that plain diverging horizontal lines were full of fine classic beauty, that arches not only made but framed pictures, that the design of a building must reckon with lights and shadows as well as forms, and that Nature must be taken into partnership and entrusted with the rich task of adding the final beauty to his work. So against the plain walls and simple arches of his designing, he plants creepers that embroider incomparable patterns over the arches and around the pillars, crowd into the corners, delicately outline windows and cornices and mass at irregular intervals along the eaves or the top of walls.

The broad sweep of green lawn seems to touch with caressing fingers the walls . . . drawing them close into the very heart of the garden. And this is done so informally and so graciously that, looking over the building and its surroundings with fresh unprejudiced mind, one could not fail to realize that the richest and most intricate carving or mad-made ornamentation of any kind would fail in beauty if compared with the decorative tracery of the green vines.

On every side the long sweeping lines of the architecture seem inspired by the gracious curve of the California landscape, fitting in admirably with the low swell of the surf and the gentle range of hills. And the result is an architectural beauty that is neither Italian nor Greek, Spanish nor French, but distinctly Californian, belonging to a new civilization with a new instinct for home building and a new belief in the value of Nature as an architectural aid.

+++

CELEBRATING SIMPLICITY IN ARCHITECTURE
BY ELOISE ROORBACH[6]

"And above all there will be no uncharacteristic or tarnished or vulgar decoration permissible, ornament being for the most part structural or necessary," says Walter Pater in his famous essay on "Style" when speaking of the literature that is to form and to maintain the literary ideal. These words apply with equal force to the architecture that is to form and maintain the architectural ideal. In fact this essay studied by literary aspirants as zealously as a theological student studies the Bible, could be read and re-read by architectural candidates to good advantage for what this distinguished analysis of pure style declares to be an essential of good literature is also found to be an architectural essential. He cautions the student of literature to "dead surplusage, as the runner on his muscles" declaring all art to "consist but in the removal of surplusage." He says that the very name ornament indicates that it is non-essential. That self-restraint, a skilled economy of means, is a great element of beauty. He warns against the use of flourishes or ornaments because of the "narcotic-force" of them "upon the negligent intelligence to which any diversion is welcome." He advises "an architectural conception of work which foresees the end in the beginning and never loses sight of it."

Whoever has read these eloquent pleas for simplicity, for a pure form of literature, will no doubt find it coming to his remembrance when they see for the first time a house recently completed in San Diego for Henry F. Timkin. The architect, Irving J. Gill, has by strict devotion to pure form, by an elimination of all ornament not "structural or necessary" created an architectural classic. He has celebrated simplicity as it is seldom celebrated in architecture, not only in this house but in much of his other work. He has created a form of architecture that attracts immediate attention because of its beauty and originality. It is true that many people unused to such severity of form do not like it. This is to be executed, for the general public has become habituated to the "narcotic" effect of over-ornamentation and object to being deprived of it.

On the other side there are many, many who see in it return to a purer art, to a more classic beauty. To them it means the arrival of a higher standard. It hints of the future hen architecture will be fine, chaste, dignified, substantial, rather than showy, ostentatious, shoddy. In the world of literature some people prefer doggerel to sonnets. In the architectural world some people peer the showy to the choice.

Mr. Gill seeing the need of a better architecture both aesthetically and practically, determined to go back to the fundamental things and start afresh from them. He began to study the fixed *principle of the line, the square, and the cube* [emphasis added]. He let them occupy his mind, absorb his thought. He experimented with them, arranged and re-arranged them. He saw that these simple things were wonderful, saw the magnitude of them, came to see their great beauty, took joy in emphasizing them. He realized that these things were the most impressive things in nature, so be began to build his houses from these models, departing as little as possible from them, accenting them in fact by using one as a foil to the other. He studied the principle of surfaces, of light and shade, of perspective, play with them composing them together over and over, until he has constructed by the laws of counterpart, as it were, an architectural fugue. (How many houses do we see that resemble an architectural ragtime?)

His idea of ornament is that it should never be fixed, conventionalized or standardized.[7] He says that this age is coming to demand standardization of all possible working things, machinery, tools, etc. But it is impossible to standardize ornament, to fix its limits, legislate its rules. So for the decoration of all his houses he plans vines and creepers just where they will grow in their own inimitable way and soften the edge of the roof, design the school at the doorway, outline an architect, etch the pillars with green, run along the wall and unite it with the house. The festoons, wreathes, corbels, arabesques, frescoes of his houses are all the artistic work of living vines and creepers. The contrast between the studied, structural severity of the house and the informal, impromptu grace of the vines, forms one of the most noteworthy features of his houses giving them a great individuality and a great beauty.

As to the construction of this house. The walls are two inches by two inches studs, placed sixteen inches on centers, with sheathing and building paper, lattice and plaster on the outside, lath and plaster on the inside. The foundations are of concrete. The outer walls are water-proofed. A special feature of this house is that the interior woodwork is finished flush with the walls in every part of the house. All doors whether leading from room to room or into the cupboards and closets, are without panels and made flush with the casings. This makes an almost dust proof home which lightens the house work to a considerable degree. . . . .

There are no corners and no projections in the whole house so that it is the acme of modern sanitary construction.

The court and the loggia floors are twelve-inch by twelve-inch red brick tile finished with a wide mortar joint. The inside walls are tinted the most delicate gray, an almost invisible gray and the ceilings chalk white. This gave all possible chance for the reflection of colors on the walls which forms such an unusual and attractive feature of all of Mr. Gill's houses. The colors of the flowers of the garden, the green of the grass, the varying tints of the sky, are reflected upon the walls of the rooms with the intensified charm of beauty and color always felt in reflections. The flame of poppies, scarlet of geraniums, blue of larkspurs, all the rainbow colors of the garden pass in turn along the walls. A bit of metal, a fold of drapery in the room becomes mirrored softly among the garden colors as the sun enters the room and touches them. . . . Because of the treatment, which is distinctly Mr. Gill's own method, there is always a soft glow in the rooms. The rooms seem tinted with delicate pastel shades one hour, become iridescent at another, are dove gray overlaid with rose at another. . . .

The large court . . . is in reality the family living room, the center of the home life. An overhead copper wire screen keeps out the flies. Vines have been planted which will soon wreath the pillars and trail from the bridge-supports of the large screen. . . . These open air rooms are coming to be the best part of the new California domestic architecture. . . .

Appendix Two

INTRODUCTION

Architect, expert renderer in pencil, and Wright's longtime friend, Chicagoan Andy Rebori was working as an editor for *Architectural Record* in New York City. One way or another, the two men agreed to an article about textile blocks. To this end in September 1927 Wright wrote penned information in a letter to Rebori.[8] Probably enclosed with the letter was illustrative material that would have supplemented that which Rebori had on hand. Rebori's article began with brief but laudatory comments about Wright's dizzy up-down career, information also gained on private social occasions during the past twenty years. The editor then mixed in snippets that reflected his friends professional troubles without specifically naming or describing them. At one point Rebori said, with characteristicCed Wrightian claims: "As is always the case where a new order aims to dislodge the old, the 'stand-patters' raise a hue and cry which is heard far and wide. Thus, Wright has made many enemies and a few ardent admirers." Included in the article were three drawings of the "Freeman, Esq." house (strangely no photos), three photographs of "The Barnsdall Residence," one of tiles imbedded in plaster on the exterior of the "Coonley Home", and one of concrete wall tiles on the exterior of Midway Gardens. The final illustration was a drawing by Richard Neutra that detailed concrete tiles for the Ennis house as previously mentioned herein.

After a plug for H. Th. Wijdeveld's 1925 book about Wright,[9] in awkward prose Rebori began reporting on the textile blocks, and that is where we begin below. Now and then other Wrightian phrases appear (such as "worthy forbears", whatever that means) and quotations conform very closely to Wright's letter. Legend, myth, fiction, confession, error, even begging, are found in this strange essay. Also, it coincides with Wright's employment as an assistant or consultant to architects for the Arizona Biltmore Hotel where his concrete blocks were the main construction material. He was otherwise out of work.

+++

FRANK LLOYD WIGHT'S TEXTILE-BLOCK SLAB CONSTRUCTION
BY A. N. REBORI[10]

> ... Later on a complete change takes place, making its appearance in his earlier California work where a monolithic mass formation is employed and that of the early types is removed from a flat roof.[11] Finally the "textile-block slab construction" is born as legitimate offspring of worthy forbears. . . . .

We are living in a scientific age of development, with the aeroplane and automobile an everyday accessory. Yet, in the general practice of architecture we are still bound to the traditional stock and trade forms of Old World buildings and ideas. Wright has succeeded in breaking the old traditions by making use of mechanical methods, modern structural forms and their application by the shaping of monolithic mass and finally by devising a method of building construction calling for the use of ornament reinforced concrete blocks which he believes will not only reduce the cost of building but which will give each architect an opportunity for individual expression along the line of a well-defined and rational building procedure. . . .

What we lack most in this country as far as building is concerned, is an exact method of work . . . making use of modern building materials at our disposal. Without such a method we cannot speed up construction, practice economy or arrive at a truly great architecture, the methods and principles of which may be universally followed.

Up to the present writing a number of experimental houses have been built with "textile-block slab construction." Since then the experimental work has been carried to a perfected state. The system consist of concrete block slabs about two or three inches thick of unit sizes which can be handled, laid on end with interlocking grooves, reinforced horizontally and vertically by means of steel rods tying the inner to the outer shell of the walls. Concrete is poured into the holes [actually the grooves] through which the rods extend, forming a complete, weatherproof, structural bond of spidery steel reinforcement between the various units making up the general system of design. The pattern or design of the face and the size of the blocks may be varied to suit any particular plan condition or exterior treatment. Floors and roof of these buildings are reinforced concrete tiles and joist knitted to the exterior and interior walls, forming a continuous construction. Walls may be made of varying reveals with the air space between the vertical slabs of the interior and outside wall face, making for warmth in the winter and coolness in the summer.[12]

Mr. Frank Lloyd Wright is now engaged in the design of a number of buildings showing the use to which his new system of construction can be applied.[13] He believes that eventually his method may be economically and efficiently used for the enclosing walls of any type of building; high, low or skyscraper.

The idea was first worked out in the construction of the Millard home—"La Miniatura"—being worked into the sunlight in this building at Pasadena during the winter of 1923.

In connection with these early experiments, it might be interesting to mention the inventor's [sic] views so we quote below from his writings on the subject.

> *"None of the advantages," says Mr. Wright, "which the system was designed to have were had in the construction of these models. We had no organization. Prepared the moulds experimentally. Picked up 'Moyana' [mañana] men in the Los Angeles street, and started them making and setting blocks—The work consequently was roughly done and wasteful.*
>
> *"None of the accuracy which is essential to economy in manufacture nor any benefit of organization was achieved in these models.*[14] *And yet the cost of the building was not more than that of a frame and stucco building of the usual Los Angeles type, of that same plan, with a good 'Spanish' exterior.*[15]
>
> *"The blocks were made of various combinations of the decayed granite and sand and gravel of the sites. The mixture was not rich. Nor was it possible to cure the blocks in sufficient moisture. The blocks might well have been of better quality.*[16]
>
> *"Some unnecessary trouble was experienced in making the buildings water-proof. All the difficulties met with were due to poor workmanship and not to the nature of the scheme.*[17]
>
> *"But it is seldom that buildings of a new type are built out-right as experimental models with less trouble than were these [?] notwithstanding our lack of organization and our concentration on* invention," *[emphasis added].*

We await with keen interest the outcome on a larger scale of Mr. Wright's latest development and contribution to the science [Rebori said] of art and building. His struggle for recognition has been a difficult one with many trials and tribulations . . . .

Once he was classed as an extreme radical and now we find that through enlightenment in modern thought and action, the "radical" is gradually being assimilated a "conservative." . . . [B]ut what we are particularly interested in is new work from a mind now mature and still vigorously capable of extending and developing a method of architecture demonstration which will be eventually understood and appreciated as it deserves.

Editor's [presumably Rebori's] Note—During the Convention of International Town Planners held in Chicago several years ago [1914], . . . the one place all foreigners wished to visit was one of Mr. Wright's buildings. Once there, they spent an entire afternoon in reverent admiration.

NOTES

1   Irving J. Gill, "The home of the future. The new architecture of the West. Small homes for a great country," *The Craftsman*, 30(May 1916), pp. 140-151, 200; reprint in Kamerling (1993), pp. 124-128.

2   Eloise Roorbach, "'Outdoor' life in California houses, as expressed in the new architecture of Irving J. Gill," *The Craftsman*, 24(July 1913), pp. 435-439. See also, idem, "The Bishop's School for Girls: a Progressive Departure from Traditional Architecture", *The Craftsman*, 26(September 1914).

3   Johann Christoph Frederich von Schiller (1759-1805), German poet, dramatist and literary theorist.

4   "Radical western architecture," *The Craftsman*, 28(August 1915), pp. 454-455. The theoretical impact on the design process of employing the geometry of square, circle and a dominant line mentioned in the two essays was explained by Gill in 1916, see note 1 above.

5   Architect Louis J. Gill, nephew and Irving's occasional partner 1914-19, was mentioned incidentally. Information for the article was based on interviews with Louis and Irving. Louis later developed a successful architectural practice in San Diego, see Hines (2000).

6   E[loise] Roorbach, "Celebrating simplicity in architecture," *WArchitect*, 19(April 1913), pp. 35-38 and plates [1-4], not reprinted herein are those parts about interior detailing or gardening or that are repetitious. Illustrated were houses for Darst (1909, demolished) and Timkin (1911, demolished) and the Mary Cossitt cottages (1910), all in San Diego. At times the text reads as if Roorbach was paraphrasing Gill. At other times it is apparent that in his writings of the 1920s Wright was paraphrasing Gill.

7   Gill's *idea* that ornament "should never be fixed . . . [or] standardized" would have appealed to Wright but "conventionalization" was a hallmark of his ornamental designs, and Gill must have known this.

8   Wright to Rebori, 15 September 1927, Wright Archive, also quoted in part in Sweeney (1994), p. 118.

9   Wijdeveld (1925). Rather than just a crafts school, Wright took the idea of a communal fellowship from H.Th. Wijdeveld, cf. Langmead/Johnson (2000).

10 A.N. Rebori, "Frank Lloyd Wright's Textile-block slab construction," *ARecord*, 62(December 1927), pp. 449-456.

11 Here Rebori refers to the pitched roofs Wright used on houses before ca.1913 as a comparison to flat roofed buildings like Midway Gardens. The propriety of flat roofs for residential buildings was controversial in Europe and North America in the 1920s.

12 This paragraph is confusing in places and erroneous in others: for example concrete was pact into grooves, not holes; neither the grooves or any other aspect of the system interlocked; it was a system of construction, not design; "joist knitted" means what?; what "varying reveals" and where are they?

13 Since Rebori believed the system had been "perfected" in 1927, he (on Wright's prompt) correctly assumed the only building under design that year was the Arizona Biltmore Hotel.

14 "These models" must refer to all four textile block houses and not Barnsdall's.

15 This sentence was not in Wright's letter. Rebori's statement that the cost to build a Wright concrete block house was equal to wood "frame and stucco" construction is quite incorrect. But was he paraphrasing Wright?

16 Only the Ennis concrete tiles had aggregates of "decayed" material from the site and they in turn decayed quickly in the concrete mix as contemporary photographs attest. Commercial sand was used for the Storer and Freeman houses, and presumably at Millard and the Barnsdall kindergarten.

17 Again, Wright blames son Lloyd (as supervisor of construction) for the problems when in fact they were the result of father's poorly designed construction system.

References

Basic sources are archival, principally the Wright Archive and the letters of his son, Lloyd, located at University of California in Los Angeles. Access to the Wright Archive was generously granted to this author by archivists Bruce Pfeiffer and Indira Berndtson beginning in the 1970s, and to the Lloyd Wright Papers by his son Eric Lloyd Wright in the 1990s.

Wright"s autobiographies, Wright (1932) and the second edition of 1943 are useful if very incomplete and often too impressionable. A third edition of 1977 released eighteen years after his death must be treated skeptically and I've assumed there were later revisers and editors.

The seminal biographical study is Twombly (1979) and the second edition is reliable on details of Wright"s life prior to about 1932. Also reliable and preferred is Secrest (1992). Other biographies or semi-biographies rely heavily on Wright's autobiographical comments or his other repetative texts or on anecdotal comments by others. They include Gill (1987) and all are of little help on matters architectural or, most noticably, on those related to community planning.

There are two bibliographies of value: Sweeney (1978) and Langmead (2003). Also of interest is <steinerag.com/flw>.

Pfeiffer (2008) contains the texts of a few books and articles all of which had been previously reprinted, often many times in whole or in part. There is a useful index but missing are the necessary illustrations as found in each text when first published. Publication and reprint information for Pfeiffer (2008) is detailed in Langmead (2003) at the date when an item was first published.

To follow are publications referred to more than once in endnotes to the text or in captions to illustrations. See List of Abbreviations at the beginning of the book.

+ + +

Aitchison, Ray. 1972. *Americans in Australia*. New York: Scribner.

Alofsin, Anthony. 1993. *FLW. The lost years, 1910-1922. A Study in Influence*. Chicago: University of Chicago Press.

------, editor. 1999. *FLW Europe and beyond*. Berkeley: University of California Press.

Atlas Portland Cement Co. 1909. *Concrete Houses & Cottages*. 2 vol. Chicago: The Company. (Another; 2 volumes in one 1910).

Banham, Reyner. 1986. *A Concrete Atlantis. U.S. industrial Buildings and European Modern Architecture 1900-1925*. Cambridge, Massachusetts: MIT Press.

Bolon, Carol R., *et al*, editors. 1984. *The Nature of FLW*. Chicago: University of Chicago Press.

Braun, Barbara. 1993. *Pre-Columbian Art and the Post-Columbian World. Ancient American Sources of Modern Art*. New York: Abrams.

Brooks, H. Allen. 1972. *The Prairie School. FLW and his Midwest Contemporaries*. Toronto: University of Toronto Press.

Brownell, Baker, and FLW. 1937. *Architecture and modern life*. New York/London: Harper. (3rd edition or printing 1938.)

Buchard, John, and Albert Bush-Brown. 1961. *The Architecture of America. A Social and Cultural History*. Boston: Little, Brown.

Chusid, Jeffrey M. 1992. "The American Discovery of Reinforced Concrete." *Rassegna* (Bologna). 49(March).

Collins, Peter. 1959. *Concrete. The Vision of a New Architecture. A Study of Auguste Perret and his Precursors*. London: Faber & Faber.; 2nd edition, Montreal: McGill-Queens Press, 2004.

Cushing, Frank H. 1882. *My Adventures in Zuni*. Palo Alto: American West, 1970. (Reprint, Palmor Lake, Colorado: Filter Press. 1967 & 1998. Reprint of articles that appeared in *Century Magazine* 1882-1883.)

Darling, Michael, et al. 2001. *The Architecture of R.M. Schindler*. Los Angeles/New York: Museum of Contemporary Art/Abrams.

De Long, David G., editor. 1996. *FLW: Designs for an American Landscape 1922-32*. New York: Abrams.

Dow, Arthur Wesley. 1899. *Composition. A Series of Exercises Selected from a New System of Art Education*. Boston: J.M. Bowles. (7th edition, Garden City: Doubleday, Page, 1910.)

Drexler, Arthur, and Thomas S. Hines. 1982. *The Architecture of Richard Neutra: From International Style to California Modern*. New York: Museum of Modern Art.

Dunham, Judith, et al. 1994. *Detail of FLW. The California Work, 1909-1974*. San Francisco: Chronicle.

Eaton, Leonard K. 2006. "Frank Lloyd Wright and the concrete slab and column." In Madge/Peckham (2006).

Encyclopedia of 20th-century Architecture. 2004. R. Stephen Sennott, editor, New York/London: Fitzroy Dearborn.

Fisher Fine Arts. 1985. *FLW: Architectural drawings and decorative art...* London: Fisher.

Ford, Edward R. 1990. *The Details of Modern Architecture*. Cambridge, Massachusetts: MIT Press.

Frampton, Kenneth. 1991. "The text-tile tectonic." In McCarter (1991). (Reprint McCarter, 2005.

Frank, Marie. 2008. "The Theory of Pure Design and American Architectural Education in the Early Twentieth Century." *JSAH*, 67(June 2008).

Freestone, Robert. 1995. "Women in the Australian town planning movement 1900-1950." *Planning Perspectives* (London), vol.10.

------. 2007. *Designing Australian Cities. Culture, Commerce and the City Beautiful, 1900-1930*. Sydney/London: University of New South Wales Press/Routledge.

------, editor 2009. *Cities, Citizens and Reform: Histories of Australian Town Planning Associations* Sydney: Sydney University Press.

------, and Bronwyn Hanna. 2008. *Florence Taylor"s Hats, Designing, Building and Editing Sydney*. Sydney: Halstead Press.

------, and Margaret Park. 2009. "Spreading the good news about town planning in Sydney 1913-34." In Freestone (2009).

Friedman, Alice T. 1992. "House Is Not a Home: Hollyhock House as 'Art-Theater Garden.'" *JSAH*. 51(September).

------. 1998. *Women and the Making of the Modern House. A Social and Architectural History*. New York: Abrams.

de Fries, Heinrich. 1926. *FLW: Aus dem Lebenswerke eines Architekten*. Berlin: Ernst Pollak.

Gebhard, David. 1971. *R M Schindler*. London: Thames & Hudson.

------. 1990. "The Myth and Power of Place: Hispanic Revivalism in the American Southwest." In Marcovich (1990).

------, editor. 1993. *The Architectural Drawings of R.M. Schindler*. 4 vols. NewYork/London: Garland.

------, and Harriette Von Breton. 1971. *Lloyd Wright Architect. 20th Century Architecture in an Organic Exhibition*. Santa Barbara: University of California.

Gelernter, Mark. 2001. *A History of American Architecture. Buildings in their cultural and Technological Context*. Hanover/Manchester, UK: University Press of New England/ University of Manchester Press.

Gill, Brendan. 1987. *Many Masks. A Life of FLW*. New York: Putnam. (Reprint, New York: Da Capo, 1998).

Gill, Irving J. 1916. "The Home of the Future. The New Architecture of the West. Small Homes for a Great Country." *The Craftsman*, 30(May).

Griffin, Dustin, editor. 2008. *The writings of Walter Burley Griffin*. Port Melbourne: Cambridge University Press.

Griffin, Marion Mahony. ca.1949. "The Magic of America." Typescript, New-York Historical Society. (Another slightly different typescript, Art Institute of Chicago, <www.artic.edu/magicofamerica/moa.html>).

Griffin, Walter Burley. 1927. "Segmental Architecture." *Australian Home Beautiful* (Melbourne), 5(1 September). (Reprint in Griffin (2008) without illustrations).

Gutheim, Frederick. editor. 1941. *FLW on Architecture: Selected Writings (1894-1940)*. New York: Duell, Sloan & Pearce. (Reprint, New York: Grosset & Dunlap, n.d.)

Hamilton, Mary Jane. 1989. [Letter.] WHistory. 73(Spring).

Handlin, David P. 1979. *The American Home. Architecture and Society, 1815-1915*. Boston: Little, Brown.

Hanks, David A. 1989. *FLW. Preserving an Architectural Heritage. Decorative Designs. . . .* New York: Dutton.

Hanna, Bronwyn. 2002. "Australia"s Early Women Architects: Milestones and Achievements." *Fabrications*. 12(1).

Haskell, Douglas. 1928. "Organic architecture: FLW." *Creative Art* (New York). No.3 (November).

Herbert, Gilbert. 1986. *Dream of the Factory-made House*. Cambridge, Massachusetts: MIT Press.

Hines, Thomas S. 1982. *Richard Neutra and the Search for Modern Architecture*. New York: Oxford University Press.

------. 1995. "Review Essay: The Search for FLW." *JSAH*. 54(December).

------. 2000. *Irving Gill and the Architecture of Reform*. New York: Monacelli.

Hitchcock, Jr., Henry-Russell. 1928. *FLW*. Paris: Cahiers d'Art.

------ [and FLW]. 1942. *In the Nature of Materials: The Buildings of FLW 1887-1941*. New York: Djuel Sloan and Pearce. (Reprint New York: da Capo, 1969).

Hoffmann, Donald. 1992. *FLW's Hollyhock House*. New York: Dover.

James, Cary. 1968. *The Imperial Hotel. FLW and the Architecture of Unity*. Rutland/Tokyo: Charles E. Tuttle.

Jandl, H. Ward. 1991. *Yesterday"s Houses of Tomorrow. Innovative American Homes 1850 to 1950*. Washington, D.C: National Trust for Historic Preservation Press.

Johnson, Donald Leslie. 1977. *The Architecture of Walter Burley Griffin*. Melbourne: Macmillan.

------. 1990. *FLW versus America. The 1930s*. Cambridge, Massachusetts: MIT Press.

------. 2004. "FLW's Community Planning." *Journal of Planning History* (Society of American City & Regional Planning History). 3(February).

------. 2005. *The Fountainheads: Wright, Rand, the FBI and Hollywood*. Jefferson, N.C/London: McFarland. (Reprint edition 2012).

------. 2012. "Design espionage: the Griffins, the Taylors, and FLW." *Fabrications*. 21(January).

------. Forthcoming. "Education and FLW."

Kamerling, Bruce. 1993. *Irving J. Gill, Architect*. San Diego: San Diego Historical Society.

Klaus, Susan L. 2002. *A Modern Arcadia, Frederick Law Olmsted Jr. & the Plan for Forest Hills Gardens*. Amhurst/Boston: University of Massachusetts Press/Library of American Landscape History.

Kruty, Paul. 1998. *FLW and Midway Gardens*. Urbana-Champaign: University of Illinois Press.

------. 2003. *Walter Burley Griffin. Architectural Models of Projects and Demolished Buildings*. Urbana-Champaign: University of Illinois.

Langmead, Donald. 2003. *FLW. A Bio-Bibliography*. Westport/London: Praeger.

------, and Donald Leslie Johnson. 2000. *Architectural Excursions. FLW, Holland and Europe*. Westport: Greenwood.

Levine, Neil. 1996. *The Architecture of FLW*. Princeton: Princeton University Press.

Lewis, Miles. 1994. "Wright, Griffin & Natco." Typescript. Paper to SAHANZ Conference, 1993, rev. 1994. University of Melbourne.

Macmillan. 1982. *Macmillan Encyclopedia of Architects*. 4 vol. New York: Macmillan.

McCarter, Robert, editor. 1991. *On and By FLW. A Primer of Architectural Principles*. 2nd edition. NewYork/London: Phaidon. (Later reprints).

McCoy, Esther. 1979. *Vienna to Los Angeles: Two Journeys. Letters of Louis Sullivan to R.M. Schindler. Letters between R.M. Schindler and Richard Neutra*. Santa Monica: Arts + Architecture Press.

------, and Randell L. Makinson. 1960. *Five California architects*. New York: Reinhold.

Madge, James, and Andrew Peckham, editors. 2006. *Narrating Architecture: A Retrospective Anthology*. London: Routledge.

Maldre, Mati, and Paul Kruty. 1996. *Walter Burley Griffin in America*. Urbana: University of Illinois Press.

Markovich, Nicholas C., Wolfgang F.W. Preiser, and Fred G. Sturm, editors. 1990. *Pueblo Style and Regional Architecture*. New York: Van Nostrand.

Meehan, Patrick J., editor. 1984. *The Master Architect. Conversations with FLW*. New York: John Wiley.

-------, editor. 1991. *FLW Remembered*. Washington, D.C.: National Trust for Historic Preservation.

Morgan, Lewis H. 1881. *Houses and House-Life of the American Aborigines*. Washington: Government Printer. Reprint, Chicago: University of Chicago Press, 1965.

Morrill, Milton Dana. 1910. "Inexpensive houses of reinforced concrete". *Architect and Engineer* (San Francisco). 21(May).

Munz, Ludwig, and Gustav Künstler. 1966. *Adolf Loos Pioneer of Modern Architecture*. New York: Praeger.

Navaretti, Peter Y. 1998. "Catalogue Raisonné." In Turnbull/Navaretti (1998).

Neutra, Richard. 1927. *Wei Baut Amerika?*. Stuttgart: Hoffmann.

Nute, Kevin. 1993. *FLW and Japan*. New York: Van Nostrand Reinhold.

Oles, James. 1993. *South of the Border: Mexico in the American Imagination 1914-1947*. Washington, D.C.: Smithsonian Institution Press.

Onderdonk, Francis S., Jr. 1928. *The Ferro Concrete Style*. New York: Architectural Book Publishing. (Reprint, Santa Monica: Hennessey + Ingalls, 1998).

Patterson, Terry L. 1994. *FLW and the Meaning of Materials*. New York: Van Nostrand Reinhold.

Pfeiffer, Bruce Brooks, editor. (xx-xx). *FLW Monograph*. 12 vol. Tokyo: Edita, 1984-88. In notes (xx-xx) refer to years discussed in a particular monograph, for example, the years 1914 to 1924 are in Pfeiffer (14-24).

------, compiler. 1984. *Letters to Architects. FLW*. Fresno: California State University Press.

------, editor. 1986. *FLW Preliminary Studies 1917-1932*. Tokyo: Edita.

------, compiler. 1992. *FLW Collected Writings. Volume 1 1894-1930*. New York: Rizzoli.

------, editor. 2002. *FLW*. Koln: Taschen.

------, editor. 2008. *The Essential FLW. Critical Writings on Architecture*. Princeton: Princeton University Press.

------, compiler. 2011. *FLW the Sketches, Plans and Drawings*. New York: Rizzoli.

Rebori, A.N. 1927. "FLW"s textile-block slab construction." *ARecord*. 62(December).

Reps, John W. 1997. *Canberra 1912*. Melbourne: Melbourne University Press.

Riley, Terrence, editor. 1994. *FLW Architect*. New York: Museum of Modern Art.

Roe, Michael. 1984. *Nine Australian Progressives*. Brisbane: University of Queensland Press.

Ross, Denman W. 1907. *A Theory of Pure Design* (Boston: the author, 1907; reprint New York: Peter Smith, 1933).

Schmidt, Henry C. 1978. "The American Intellectual Discovery of Mexico in the 1920s." *South Atlantic Quarterly*. 77(3).

Schwarzer, Mitchell. 2004. "Adolf Loos". In *Encyclopedia*.

Scully, Jr., Vincent. 1960. *FLW*. New York: George Braziller.

------. 1969. *American Architecture and Urbanism*. New York: Praeger.

------. 1988. Introduction to Bolon (1988).

Secrest, Meryle. 1992. *FLW: A Biography*. New York: Knopf. (Reprint Chicago: University of Chicago Press, 1998).

Siry, Joseph M. 1996. *Unity Temple. FLW and Architecture for Liberal Religion*. New York: Cambridge University Press.

------. 2008. "The Architecture of Earthquake Resistance. Julius Kahn's Truscon Company and FLW's Imperial Hotel." *JSAH*, 67(March).

Slaton, Amy E. 2001. *Reinforced Concrete and the Modernization of American Building 1900-1930*. Baltimore: Johns Hopkins University Press.

Sloan, Maurice M. 1912. *The Concrete House and Its Construction*. Philadelphia: Association of American Portland Cement Manufacturers.

Smith, Elizabeth A.T., and Michael Darling. 2001. *The Architecture of R.M. Schindler*. Los Angeles: Museum of Contemporary Art.

Smith, Kathryn. 1985. "FLW and the Imperial Hotel. A Postscript." *Art Bulletin* (New York). 67(June).

------. 1992. *FLW. Hollyhock House and Olive Hill. Buildings and projects for Aline Barnsdall*. New York: Rizzoli.

------. 2005. "The L.A. Textile Block Houses." *FLWQ*, 6(Summer).

Spathopoulos, Wanda. 2007. *The Crag. Castlecrag 1924-1938*. Blackheath, NSW: Brandl & Sclesinger.

Spence, Lewis. ca.1928. *The Mysteries of Britain*. 4th ed. London: Rider.

Sprague, Paul E., editor. 1990. *FLW and Madison: Eight Decades of Artistic and Social Interaction*. Madison: Elvehjem Museum of Art.

Starr, Kevin. 1990. *Material Dreams. Southern California Through the 1920s*. New York: Oxford University Press.

Steele, James. 1992. *Barnsdall House FLW*. London: Phaidon.

Stephen, Ann, Andrew McNamara and Philip Goad. 2006. *Modernism & Australia. Documents on Art, Design and Architecture 1917-1967*. Melbourne: Miegunyah Press.

Stephens, John L. 1843. *Incidents of Travel in Central America, Chiapas, and Yucatan*. Frederick Catherwood, illus. 2 vol. New York: Harper & Brothers.

Storrer, William Allin. 1993. *The FLW Companion*. Chicago: University of Chicago Press.

Sullivan, Louis. 1924. "Concerning the Imperial Hotel, Tokyo, Japan." *ARecord*, 55(February). (Reprint, idem, "Concerning the Imperial Hotel, Japan" in Wijdeveld, 1925).

Sweeney, Robert L. 1978. *FLW An Annotated Bibliogaraphy*. Los Angeles: Hennessey & Ingalls.

------. 1994. *Wright in Hollywood. Visions of a New Architecture*. Cambridge, Massachusetts: MIT Press.

------, and Charles M. Calvo. 1984. "FLW: The textile block houses." *Space Design* (Tokyo), 7-9, no. 240(September).

Tanigawa, Masami. 1980. *Measured Drawing FLW in Japan*. Tokyo: Tokodo Shoten.

Taylor, George A. 1914a. *Town Planning for Australia*. Sydney: Building.

------. 1914b. "A speak-up for Australia." *Building*. 15(12 October).

------. 1915. "There." *Building*. 15(12 July).

------. 1916. *"There!" A Pilgrimage of Pleasure*. Sydney: Building.

------. 1924. *The Ways of the World*. Sydney: Building.

Thomas, David Hurst. 2000. *Skull Wars. Kennewick Man, Archaeology, and the Battle for Native American Identity*. New York: Basic Books.

Thomson, Iain. 1999. *FLW A visual encyclopedia*. London: PRC.

Toker, Franklin. 2003. *Fallingwater Rising: FLW, E.J. Kaufmann, and America's Most Extraordinary House*. New York: Knopf.

Torrence, Bruce T. 1982. *Hollywood: The First Hundred Years*. New York: Zoetrope.

Treasures of Taliesin. Seventy-six Unbuilt Designs. Bruce Brooks Pfeiffer, editor. Fresno/Carbondale: California State University Press/Southern Illinois University Press. 1985.

Trost & Trost. 1924. "A Hotel in Pueblo Architecture." *ARecord* (June 1924).

Tselos, Dimitri. 1953. "Exotic Influences in FLW." *Magazine of Art* (New York). 47(April).

Turnbull, Jeff, and Peter Y. Navaretti, editors. 1998. *The Griffins in Australia and India. The complete work and projects. . . .* Melbourne: Miegunyah Press.

Twombly, Robert. 1979. *FLW His Life and His Architecture*. 2nd edition. New York: John Wiley.

Vernon, Christopher. 1996. "Berlage in America: The Prairie School a 'The New American Architecture.'" In Jan Molema, editor, *The New Movement in the Netherlands 1924-1936*. Rotterdam: 010, 1996.

------. 1998. "An 'Accidental' Australian: Walter Burley Griffin''s Australian-American Landscape Art." In Turnbull/Navaretti (1998).

Viladas, Pilar. 1985. "Invisible Reweaving." *PArch*. 66(November).

Viollet-le-Duc, Eugene. 1876. *The Habitations of Man in all Ages*. Benjamin Bucknall, trans. London. (Reprint New York: Benjamin Blom, 1971).

Waldheim, Charles, and Katerina Rüedi Ray. 2003. *Chicago Architecture: Histories, Revisions, Alternatives*. Chicago: University of Chicago Press.

Watson, Anne, editor. 1998. *Beyond Architecture. Marion Mahony and Walter Burley Griffin. America, Australia, India*. Sydney: Power House Museum.

Weintraub, Alan, and Thomas S. Hines, 1998. *Lloyd Wright. The Architecture of FLW Jr.* London: Thames & Hudson.

------, and Alan Hess. 2007. *FLW Prairie Houses*. NewYork: Rizzoli. 2006.

Wermiel, Sara E. 2000. *The Fireproof Building. Baltimore*. Baltimore: Johns Hopkins University Press.

Whipple, Harvey, editor. 1920. *Concrete Houses. How They Were Built*. 2nd edition. Detroit: Concrete-Cement Age Publishing.

Wijdeveld, H.Th., ed. 1925. *The Life-Work of the American Architect FLW*. Santpoort, Nederlands: C.A. Mees.

Wright, Frank Lloyd. 1908. "In the cause of architecture." *ARecord*. 23(March).

------. 1910. *Studies and Executed Buildings by FLW*. Palos Park, Illinois: Prairie School Press, 1975. Reprint of *Ausgefurte bauten und entwurfe von FLW*, (Berlin: Ernst Wasmuth, May 1910), and as *Studies and Executed Buildings by FLW*, (Chicago: Ralph Fletcher Seymour, June 1910), English text. Both English and German text are found in the Palos Park edition.

------. 1911. *FLW: Ausgefürte Bauten*. Berlin: Ernst Wasmuth. (Reprint as *FLW The early work*, New York: Horizon, 1968, and as *The early work of FLW. The "Ausgeführte Bauten" of 1911*, [New York: Dover, 1982]. Variant, *FLW, Chicago*. (8. Sonderheft der Architektur des XX Jahrunderts.) Berlin: Ernst Wasmuth, 1911.

------. 1914. "In the cause of architecture. Second paper." *ARecord*. 35(May).

------. 1923a. "In the Cause of Architecture. The New Imperial at Tokyo." *WArchitect*, 32(April).

------. 1923b. *Experimenting with Human Lives*. Pamphlet. Los Angeles: Fine Arts Press. (Also Chicago: Ralph Fletcher Seymour, 1923).

------. 1923c. "In the cause of architecture. In the wake of the quake concerning the Imperial Hotel, Tokio." *WArchitect*, 32(November).

------. 1924. "In the cause of architecture. In the wake of the quake." *WArchitect*, 33(February 1924).

------. 1925. "In the cause of architecture. The third dimension." *Wendingen* (Amsterdam), 7(4). (Reprint in Wijdeveld,1925).

------. 1927a. "In the cause of architecture. II. Standardization, The soul of the Machine." ARecord. 62(June).

------. 1927b. "In the cause of architecture. IV. Fabrication and Imagination." *ARecord*. 62(October).

------. 1927c. "Why the Japanese Earthquake Did Not Destroy the Hotel Imperial." *Liberty* (New York). 4(3 December).

------. 1928. "In the cause of architecture. VII. The meaning of materials—concrete." *ARecord* 62(August).

------. 1929. "A Building Adventure in Modernism. A successful adventure in concrete." (American) *Country Life*. 56(May).

------. 1929a. "Surfaces, and mass,—again." *ARecord*. 66(July).

------. [1931]. *Two Lectures on Architecture*. Chicago: Art Institute.

------. 1932. *An Autobiography*. New York/London: Longmans Green. (Reprinted and condensed in Bruce Brooks Pfeiffer, ed., *FLW collected writings*, volume 2, [New York:

Rizzoli, 1992]; reprint Petaluma, California: Pomegranate Communications, 2005; and Pfeiffer (2008), each with different pagination and fewer illustrations.

------. 1943. *An Autobiography.* New York: Duel Sloan & Pearce.

------. 1957. *A Testament.* New York: Horizon.

Wright, John Lloyd. 1946. *My father Who is on Earth.* New York: Putnam's. (Reprint New York: Dover, 1992).

Wright Studies. 1992. "Taliesin 1911-1914." Vol. 1. Carbondale: Southern Illinois University Press.

About the Author

Donald Leslie Johnson taught at Flinders University in Adelaide, South Australia from 1972 until retirement in 1988. He is now an adjunct professor of architectural history at the University of South Australia. Previously he practiced architecture in Seattle, Philadelphia, and Tucson and taught theory and design at Arizona, Washington State and Adelaide universities. He has written extensively in books and articles about the architecture of Frank Lloyd Wright, Walter Burley Griffin and Canberra, Australian architectural history, and on American and Australian city planning history. Currently he is completing research on Wright's education and on the matrix that composed critical thinking about the need for urban parks two centuries ago. Johnson was in Louis I. Kahn's masters class of 1960-1961 at the University of Pennsylvania. He was a member of the American Institute of Architects and is an Honorary Fellow of the Royal Australian Institute of Architects.

Index

Numbers in **bold** indicate a page with illustrations.
FLW = Frank Lloyd Wright

Academy of Arts, Vienna 165
Adler, Dankmar 56, 103, 137
Adler and Sullivan, architects xx, 7, 103, 121, 131, 149, 158, 184
adobe xiii, 30n69, 81, 105, 114, 121, 127, 144-147, 149, 155, 159, 182
  Actor's Adobe 154
  concrete adobe 163-165, 187, 192-193
  pre-Columbian 125
  and Sachs house 186
aesthetics 19, 22, 85-88, 99, 110-112, 191-193
Albright, Harrison 6, 148
Allen house (Gill) 6
Allison, James E. and David C., architects 184
Alofsin, Anthony 2, 114
American Concrete Institute 16
American Contractor 17
American Institute of Architects 86
Amsterdam, Netherlands 182
Andrè, Jacques 96n23
architectonics xxi, xxii, 3-4, 47 123
Architectural League of America 87
Architectural Record 55, 57, 89, 104, 105, 182, 203
Arizona xi, xvi, 43, 127, 144, 146, 165, 168n5, 187, 190, 191
Arizona Biltmore Hotel 12, 25, 48, 90, 105, 114, 193, 203, 206n13
Arizona Historical Society, xi, 7
Armstrong, Jessie 185
Armstrong, William Henry 185
"Art and Craft of the Machine" 87, 182

Art Institute of Chicago 55, 85-87, 148
artificial stone 13
Arts and Crafts /Movement, English/ American 2, 16, 22, 32, 54, 57, 85-86, 105, 120, 130, 133, 143, 145, 147, 159
  German 20
Aspdin, Joseph 11
Atlas Portland Cement Company 17
Atterbury, Grosvenor viii, xii, 6, 12, 18-21, 100, 105
  Forest Hills Gardens, 20
  nailcrete 24
  Standardized Sectional System viii, 19, 113
  workman's cottages 19-20
Australia xv-xvi, xxii, 6, 7, 10, 25, 53-67, 9473-74, 79, 127, 179-181
  Parliament House competition 1914 7, 73
  — the Griffins 73
Aztec buildings xxi, 61, 120, 124-126, 134, 136

Banham, Reyner, engineer/architect 16
Barker, Nelson E house xii, 6, **148**
Barnsdall, Aline 2, 3, 32, 33, 35, 48, 152, 161, 163, 165, 186, 203
  artists' colony 154
  — hollyhocks 126
  Director's House 159
  hermaphroditic 192
  house/Hollyhock vii, vii, xiii, xxi, 2-4, **4, 5**, 8, 9, 39, 46, 79, 94, 123, 124, 127, 129, 136, 137, 148, 153-157, **155**, house Beverly Hills 9, 83, 126, 129, 185
  kindergarten vii, ix, 9, 35, 41, 43-45, **45**, 46, 206n16
  — aberration 194

— construction stopped 9, 44, 90
— and Pauline Schindler 44
pueblo 152
Residence A vii, 167
Residence B vii, 41, 44, 126, 159, 171, 173
terrace houses 102
theater xii, 123 149, 154, 156
as socialist 3, 163
Barnsdall Art Park 3
Barton, Francis M., engineer 14
Bateman, Joseph R. viii
Baudot, Anatole de, architect 15
Bay City, Michigan 113
Beauty and Utility in Concrete 18
Beijing, China 8, 61
Bella Vista Terrace houses 6, 147
Berlage, Hendrik Petrus, architect 156, 162, 186
Berlin, Germany 6, 56, 152, 161, 181, 182
Berndtson, Indira xv
Beverly, Chicago 80
Beverly, Massachusetts 5, 11
Beverly Hills, Los Angeles 9, 83, 126, 127, 187
Beye Boathouse project 164
Big Tree Wigwam 160
Billson, Edward F. and knitlock 95n5
Bitter Root Valley, Montana 134
Blackburne, E.L., architect 13
Blackhawk Country Club 160
Boari, Adamo, architect 191
Bogk, Frederick C. house xii, 7, 33, 132-**133**
Bollman house 8, 80-81, 130, 159
Boston, Massachusetts 87, 148, 189
Boy's Dormitory, Childrens' Home Association 6
brick 11, xiv, 18, 21, 24, 47, 61, 63, 80, 92, 102, 107-108, 110, 157
   Atterbury 20
   concrete 11, 14
   Foster house 186
   German warehouse 131-132
   Medieval construction 113
   Midway Gardens 112
   pierced brick, St. Jean 15
   Roman 10
   and Gill 102
   see clay, terra cotta
Brigham, Edmund D. house 17
Britain *see* United Kingdom

British Empire Exhibition 180
Brodie, Alexander, engineer 18, 19, 21
Brown, August O. xv, 35
building blocks, "patent mold for", 1850 12
Buffalo, New York xiv
Burchard, John 1, 187
Bush-Brown, Albert, architect 1, 187
Byrne, Francis (Frank) Barry, architect 7, 55, 160, 191
   and the Griffins 58, 66, 127

Canada 55, 58, 73, 179, 192
Canberra xxii
Carnegie Steel Corporation 20
Carrère and Hastings, architects 14
cast stone 13
Castlecrag, New South Wales 179-180
   Anthroposophy/Theosophy 180
cement (Portland) 3, 10, 24, 39, 83, 92, 94, 105, 106, 112, 113
   history 11-18
   stucco xii, xiii, 3, 12, 13, 14, 18, 23, 28n41, 43, 73, 79-81, 102, 125-127, 129, 130, 150, 153, 205, 206n15
Cement Age 16
Cement and Engineering News 17
Cement Era 17
Cement Houses and How to Build Them... perspective views and floor plans of concrete block... 17
Cement World 17
Century 158
Chandler, Arizona 105, 187, 190
Chappaqua, New York 13
Cheap Cottage Exposition 18
Cheney, Charles H. 180
Cheney, Mamah xx, 7, 64, 161
Cheney, Sheldon 156
Chicago 2, 32, 165
Chicago Architectural Club 55, 85, 87
Chicago Herald-American 180
Chicago Home Prize Competition 180
Chicago Portland Cement Company 18
Chicago Tribune 180
China 8, 61, 134, 161
Chusid, Jeffrey M. xv
City Club Bulletin 182
clay 43, 74, 79, 80, 85, 100, 107
   Barnsdall theatre stage xii, **155**
   clay/terra cotta 80
   hollow blocks 3, 43, 81, 93, 108, 150

Imperial Hotel 103
and cement 11
see brick, terra cotta
Clèrisseau, C.L. 137
Cliff Dwellers of Mesa Verde 158
Coignet, Francois 11
Collins, Julie xv
Colter, Mary, architect 158
Columbus, New Mexico 7
communist 3, 162, 184
Composition 88
concrete 10
  aesthetics 14-15
  history 10-18
  hydraulic 10, 11, 17-18
    — ancient 10
  mixture 11-12
  painted floor, GIII 92
  panels, wall/roof xiii, 6, 23, 113
    — Atterbury's 6, 18-21, 113
    — Brodie's 19
    — Conzelman's 23
    — Schindler's 13, 150
    — tilt up 150
  pozzuolana 10
  precast vii, ix, xi, xii, 1, 6, 10, 16, 18-21, 24, 31
    — hollyhocks, etc 116, 126, 131, 132-133, 192
    — patents 14
    — Unity Temple 15-16
  reinforcement, metal 13, 14
  "sectional mold", Morrill 21
  ships 17
  slip form 81, 177, 182
  tamping 12, 47, 83
  tilt-up walls 176
Concrete 17
concrete block
  cinder aggregate 19, 24
  Florida Southern College 92
  machines 24
  molds, moulds xi, 20, 47, 83, 105, 164
    — glass 85
    — metal 14, 19, 21, 23, 74, 90
    — wood 12, 16, 18, 90, 110, 113
  patent drawing, FLW x, 81, **82**, 83, 94
    — shapes **93**
  plagiarism 90
  problems FLW 89-94, 113-114
  wall tile system FLW 90-**91**
  water infiltration 92

Concrete Block Houses 17
Concrete Block Machine Manufacturers Association 16
Concrete-Block Manufacture; Process and Machines 17
Concrete Builder 17
Concrete City viii, 6
Concrete Engineering 17
Concrete Houses and Cottages 17
Concrete Houses and How They Were Built 18
Concrete Tenement Dwelling 19
Conzelman, John E., engineer 23
Coonley house 110
Corwin, Cecil, architect 191
Cotè, Duane, architect xv
Craftsman, The 120, 199
creativity 1-2
Creighton, D.F., architect 22
Cushing, Frank H. 158, 169n55
Custer, General George 158

Dana house 2
Dean, George, architect 86, 191
Death Valley, California 9, 186, 187
Deep Springs, California 186
  Deep Springs College 186
Delaware, Lackawanna and Western Railroad Truesdale Colliery 21
design process 1-4, 121, 126, 194, 206n4
  and art 152
de Swart, Jack xv
Dictionary of Architecture 12
Dodd, William J., architect 184
Dodge, Walter L., house 7, 151
Doheny, Edward L. 9, 33, 163, 184-185
Doheny, Estelle 33
Doheny Library, University of Southern California 34
Donnelley & Sons, R.R. 17
"double shell" construction xi, 108, **109**
Dow, Arthur Wesley x, 87
  two dimensional design 87-89, **89**
Doyle, Arthur Conan 3
Drummond, William, architect 191
Dubrucq, Virginia 143
Dune house project 159
Dunning, Max, architect 85
Duquesne, Eugene 21, 29n53
Duse, Eleanor 156

Eagle Rock, California xii, 8, **153**

earthquake xi, 39, 113
  Kanto/Japan 9, 100, 102-103
  Los Angeles 41, 45, 51n36, 83, 92
École des beaux arts 21, 86, 87, 143
Edison, Thomas A. 14
  Edison Portland Cement Company 14
    — "Poured Cement House" 14
El Pueblo Ribera Court xiii, xiv, 9, 177, **178, 179**, 181
Elgh, George 79
Elk Hills, California 185
Elzmer and Anderson, architects 11
Emerald Bay, Lake Tahoe 185
Emerson, Ralph Waldo xx, 31, 120, 137, 143
Endo, Arata 104
Engineering and news 17
Engineering World 17
Ennis, Charles and Mabel 32, 39, 163, 186
  aesthetics 85, 124, 136, 163-164, 192-193
    — concrete tiles 85, 89, 91, 93, 98, 206n16
  house ix, x, xv, ix, 35, **42, 43**, 46-47, 82, 129, 133, 172
    — construction problems 41, 51, 92, 104-105
  and the Griffins 66
Erie Canal 17
Euroamerican 133, 157-158, 160, 184, 185
Europe 6, 7, 10, 18-19, 32, 43, 55, 74, 81, 100, 151, 178-180
  modernism 23, 25, 31, 107, 119-120, 122, 133, 134, 147, 152, 158, 178
  war 58, 62, 63, 184, 188-194
  and FLW 64, 132, 137, 148, 156, 197n29
  and Neutra 181
  and Sullivan 161
Experimenting with Human Lives 9

faience, definition 197n45
Fall, Albert 185
Fallingwater house 1
Fallingwater Rising 2
Fanette Island, Lake Tahoe 185
Federal Bureau of Investigation 3
Feller, A 197n29
Fenollosa, Ernest 87
fiction 99ff
Fisher Fine Art xiv

Flagler College 14
Florida 2, 15, 178
Florida Southern College 2, 79, 90, 92
Forest Hills Gardens 20-21
Foster, Dorothy Martin house xiv, 9, **187**
  aesthetics 186-187
Foster, James 186
France 6, 23, 63, 71, 96, 119, 184
Frank, Marie 85
FLW and Madison: Eight decades of artistic and social interaction 1
FLW Foundation xv, xvi
FLW versus America: the 1930s 2
Frankston, Victoria 74
Frear, George A., Chicago 13
  Frear Stone 13
    — H.B. Horton house, Chicago 13
Freeman, Samuel and Harriet 3, 9, 31, 35, 39, 41, 66, 76
  house ix, x, xv, 35, **38, 40, 41**, 44, 46, 83, 164, 187, 203
    — additions 163
    — construction problems 41, 88, 92, 112
    — and Walter Griffin 180
Freestone, Robert xv
French, William, landscape architect 86
Fries, Heinrich de, architect 181
Fritz, Sr, Herbert, architect xi
Fuermann, Henry vii, xi

Gaardboe, Mads xv
Garden, Hugh, architect 191
Garden City 65
  Letchworth 18
Garnauat, Christine xv
Garnier, Tony, architect 21, 151
Gazette des Beaux-Arts 181
Gebhard, David xv, 81, 119, 144
Gelernter, Mark 143
George M. Millard Rare & Fine Imported Books 33
German, Albert Dell 7
  warehouse x, xii, 7, 14, 51n36, 88, 110, 113, **131**, 132
    — aesthetics 124-125, 131-133
Germany 10, 63, 122, 151, 181-182
  Garden City 20, 68n10
  Mendelsohn 160
  and FLW 6, 161
  and the Griffins 181
  and Neutra 90

Giebner, Robert C. ix, xiii, 182
Gilbert, Cass 119
Gill, Brendan 123, 125
Gill, Irving xii, xiii, xxii-xxiii, 6-8, 143, **148**, 151, 200-206
  aesthetics 193-194, 199
  cube house xii, **150**
  Irving Gill Distinguished Architect Award 169n26
  regionalism **156**-157
  and Adler and Sullivan 121, 149
  and FLW 159, 164, 175, 184
  and Frank Mead 152
  and John L. Wright 149
  and Lloyd Wright 81, 146-147, 149
  and Morrill 23, 178
  and North Africa 182
  and Schindler 165, 175-178
  and Silsbee 149, 184
  and the southwest 144-146, 161, 164, 191-192
Gladney, Edna Browning house 160, 171n61
Glen Alden Coal Company 21
Glencoe, Illinois 17
Goodhue, Bertram, architect 149
Gothic 12, 15, 16, 119, 125, 127, 133, 161, 196
Graf, Otto A. 123
Greeley, Horace 13
Greene brothers, architects 191
Greenough, Horatio 31
Gregg, Harry house 145
Grey, Elmer, architect 184
Griffin, Genevieve 59
Griffin, George 7
Griffin, Walter Burley and Marion (Lucy) Mahony Griffin x, xi, xvi, 6, 7, 10, 25, 63, 65-67, 71, 74, **75**, **128**, 167, 180
  early years 53-56, 73, 80, 86
  — Pholiota house x, **75**
  Canberra, Direction of Design and Construction 6, 60
  — abolished 63
  Lucknow, United Provinces, India 179-180
  Natco 79
  Purcell 64
  USA 1925-27 94
  and FLW's textile blocks 180-181
  and Neutra 181
  and the Taylors 59-61, 64-66

and white australia 62
  see also knitlock; Taylor, George; Canberra; University of New Mexico
Groff, Gerald and Marian xv
gunite concrete 14, 182

Hanna, Bronwyn xv
Harding, Warren G., President 185
Hardy, Thomas Paul xi, 100, 115n2
  houses 100, **101**
  see also Monolith, Schindler
Harper's Magazine 158
Hart, Russell, architect 24
Harvey, Fred, restaurants 158
Haskell, Douglas 134, 164, 190
Hebbard, William S. 145
Heidelberg, Victoria x, 74
Hennebique, François, architect/builder 14
Hickox, Warren, house xi, **130**, 190
Highland Park, Illinois xii
Hillore, Q.A. 13
Hillside Home School 148, 159
Hines, Thomas S. xiv, 123, 146-147, 150, 194
Hinzenberg, Olgivanna 9
"Hispano-Moresque aura" 134
Historic American Buildings Survey 182
Hitchcock Jr., Henry-Russell 182
Hoffman, Donald 2
Hoffman, Julius 181
Holladay, Ben 185
Holland 20, 154
Hollow Concrete Block Building Construction 17
hollow clay blocks/masonry 3, 43, 79, 80-81, 93, 108, 150, 155
Hollyhock house see Barnsdall, Aline
hollyhocks see Barnsdall, Aline
Hollywood Art Association 163
Hollywood Bowl 94, 159, 171
Hollywood, California vii, ix, xii, 2-4, 7-8, 35, 38, 39, 45, 48, 80, 133, 149, 155, 163-164, 183
Honolulu, Hawaii 13
  Honolulu Post Office 13
Hopi 158
  legend 175, 194
Horatio West Court xiii, 8, **156**, 157, 176-177
House Beautiful 17
  "The Age of Concrete" 17

house kits
　Gordon Van Tine, Montgomery War,
　　Bennett, Sears, Roebuck, Aladdin
　　113
How, James Eades house 176, 178, 195n3
*How to Manufacture Concrete Hollow
　Blocks* 17
Hull-House, Chicago 44, 182
Hunt, Myron, architect 86, 184, 191
Huxtable, Ada Louise 136
Hyman, Elaine *see* Kira Markham

Idaho 158
imitation 3-4
　borrowings 2, 4, 190
Imperial Hotel, Tokyo x, xii, 2, 6, 7, 85, 88,
　　108, 132, 148, 162, 165
　aesthetics 125, 126-127
　regressively effusive 192
　structure/earthquake 100, 103-104,
　　109
　and Louis Mullgardt 61
Important Events, 1903-1925 5-9
in situ concrete vii, xi, xii, 6, 15-16, 18, 21,
　　48, 73
　Imperial Hotel 108
Indians 121, 133-134, 153, 157, 159-160
　Algonquin 159
　Blackhawk 160
　Chippewa 160
　Chumas 153
　Lakota 158
　Pueblo 154, 156
　　— Zuni 154, 158
　southwest 133, 137, 144, 147, 164
　Winnebagos 259
　as childs-play 160
Ingalls (or Transit) Building, Cincinnati,
　　Ohio 11
Inland Architect 17
Innelli, Alfonso xi
International Studio 137
International Style 122
　FLW's opposition 122
invention 1-3
Italy 6, 161
　Renaissance 120

Jacobs Jr., John M. 136
Japan 2, 6
　two dimensional art 2, 86-87
Jenkins, David Charles x, 74, 95

Johnson, A.M. 9, 126, 185
　Death Valley 9
Johnson, Adam Kaare xv
Johnson, Donald Leslie, architect 2, 123,
　　217
Johnson Wax building 83, 160
Jones, Lloyd, family 2
Jones, Owen
　on decoration 192
Jones, Rev Jenkin Lloyd 120, 138n4, 144

Kahn, Albert, architect 14
Kahn, Julius, engineer 14
Kahn, Louis Isador, architect 137
Kankakee, Illinois xi, 130
Karma house 152
Kaufmann, Edgar 2
　Fallingwater house 2
kindergarten, see Barnsdall
knit-block viii, 73, 79, 125, 159, 167,
　　193
　and knitlock 80-81
knitlock (construction system) x, 7, 25,
　　65-67, 73-79, **75, 77, 78**, 81, 83, 99,
　　112, 167, 180-182, 193
　compared to FLW's block **84**, 94
　and knit-block 80-82, 193
　and Marion Griffin 180
　and Richard Neutra 10
　as Segmental Architecture 74, 76
Korab, Balthazar ix
Korab, Christian ix
Kruty, Paul xvi, 2

La Jolla, California xiii, xiv, 7, 9, 150, 177-
　　178, 181
La Miniatura, see Millard, Alice house
Lacelles, W.H. 11, 13-14
Lake Geneva, Switzerland 162
Lake Tahoe 9, 160, 163, 184
　Emerald Bay 185
　sculptures 160
Lalous, Victor 63
Lambot, Jean-Louis, boat hulls 11
landscape vii, viii, xxiii, 3, 54, 60, 86, 144,
　　146, 159, 179, 180-181, 187-188, 189,
　　200
　Barnsdall house 3
　Chinese 184
　Forest Hills 20
　Griffin 54
　southern California 146

landscape architect 20, 21, 81, 86, 148, 149, 165
Langmead, Donald xv
L'Architecture 181
Larkin Administration Building x, 58, 73, 163, 164, **188**, 186, 192
　demolished 1950 xiv
Le Corbusier 151, 182
League of Nations 180
Leeds, England 11
Lehigh Valley cement plant, Pennsylvania 11
Lentz, Jeffrey B. ix, xiii
Letchworth, UK 18
Levine, Neil vii, xii, 114, 133
Levine, Robert house 92
Lewis, Miles xv, 94
Liberty magazine xi, **109**
Lippincot, Roy 59, 62
liquid stone 13
Liverpool, UK 18-19
Lloyd Jones cemetery 2
Lloyd, Dr. W. Llewellyn 177
Lockard, Kirby xv
Lockington, W.N., architect 13
Long, Birch, architect 191
Loos, Adolph 151
　"Ornament and crime" (*Ornament und Verbrechen*) 151
　Austrian culture 151-152
　Wright's Wasmuth folio 152
　in U.S. 151-152
Lorch, Emil, architect 86-88
Los Angeles vii, viii, ix, xiv, xxi, xx, xxi, 2, 3, 4, 31-36, 39, 41-43, 55, 58, 64, 66, 74, 81, 88, 90, 94, 102, 104, 125, 137, 163-165, 181-187, 193-194
　Mexican labor 110, 205
　Municipal Art Gallery vii, 3
　and Gill and Lloyd Wright 149, 151, 175
　and the Griffins 180
　and Schindler 100
　see also Barnsdall
Los Feliz vii, 3
Lovell, Philip or Lean 3, 43, 196n19
　Neutra house xiv, 10, 178, 182, **183**, 186, 193
Lowes, G.P. house xii, xiv, 8, **153**, 165
　Schindler 163, 170
Lowry, T.J. 12
Luxfer Prism Company 85
Lucknow, India 179

MacCormac, Richard 123
McArthur 114
McCarter, Robert 123
McCarthy, Cormac 175
McCormick house project 164, 165
McKim, Mead and White 134
Madison, Wisconsin xx, 1, 2, 9, 123, 148, 159, 160, 184, 185, 187
Madison Realty Company 159
Madison Square Garden, New York 18, 71
Mahony, Marion Lucy *see* Griffin, Marion
Malvern, Victoria 74
Manning and Macneille, architects 14
Manufacture of Concrete Blocks and Their Use in Building Construction 17
Markham, Kira 162-163
Martin, Darwin D. 186, 192
Martin, T.P. house xiii, 7, **166, 167**
　and Barnsdall house 165
Massachusetts 5, 11
Massachusetts Institute of Technology 54, 86
Maya, Yucatan, Mexico xxi, 154
　buildings xii, 119-120, 123, **124**-126, 132-134, 136-137, 190-191
　— Puuc Maya (Chicago) 158
　— Chichèn Itzà 131
　Governor's House xi
　see Uxmal, Mexico
Maybeck, Bernard, architect 191
medieval laborer 109
Melbourne, Victoria x, xv, 7, 24-25, 55, 59, 89, 65, 79, 95
　Herald 65
　Royal Melbourne Show 66, 74
Mendelsohn, Erich (Eric) 160
　and Neutra 181
Mexican labor 108
Mexico xii, 6, 8, 114, 119, 122, 124, 172, 189
Michaelerplatz Building 152
Midway Gardens, Chicago xi, 2, 6, 18, 56, 88, 97n29, 110-112, **111**, 113, 116n38, 119
　aesthetics 125
　demolition 116n38
　and John L. Wright 148
Mill Run, Pennsylvania 2
Millard, Alice viii, 31-33, 39, 66
　house Pasadena vii, ix, x, **33, 34, 36**, 44, **46**, 53, 65, 79, 80-83, 85, 125, 204
　— addition to 163

— concrete block face ix, 164, 192
— construction problems 91-92, 110, 113
— exotic sources 133-136
— interior **36**
— and southwest 154, 161
house Highland Park xii, 32, 130, 153, **154**, 190
Millard, George Madison xii,
Millard, Mrs George M. *see* Millard, Alice
Millet, Louis J. 86-87
Mitla, Mexico 124
Miltmore house (Gill) 6
Milwaukee, Wisconsin xii, 7, 33, 132, 133
Mississippi valley 119
Monier, Joseph 11
Monolith Homes xi, 100, **101**, 102, 104, 165
see also Hardy
Montreal, Canada 192
Montana 158
Monterey 158
Mook, Robert, architect 13
Morrell, J.C. 55
Morrill, Milton Dana viii, xxii, 6, 21-23, 100, 113, 122, 143, 151-152, 178
building sites, cities 22
Moser, Werner 197n29
Mound Builders 119
Mount Rainier Nation Park 160
Mullgardt, Louis 61
Mumford, Lewis 194
Munoz, Oskar xv
Musèe des Arts Dècoratifs 181
"My Adventures in Zuni" 158

Nahuan Palace, Yucatan xii, **135**
Nailcrete 24
Nakoma Country Club 9, 159-160, 163, 185
Nancy, France 96
Nanticoke, Pennsylvania viii
Natco Company 8, 25, 73, 74, 79-80, 81, 108, 155
tex-tiles, textured tiles 79
National Conference, on Concrete House Construction 18
National Library of Australia
Eric Nicholls Collection xi
National Life Insurance Company building 186
Navaretti, Peter xv, 74

negro 55, 62
Confederate states 184
Nel-Stone xi, 94, **93**
Netherlands *see* Holland
Neutra, Richard x, xiv, 3, 9, 10, 43, 175, 180-182, 193, 197n29
How house 195n3
Lovell house **183,** 186
wall tile system FLW 90-**91**
and the Griffins 181-182
and Rudolf Schindler 66, 176, 178
and Hollywood 163
New England 85
New South Wales, Australia xvi, 53, 59
New World 120
New York, State/City xvi, 6, 9, 11, 13, 14, 17, 18, 19, 22, 55, 58, 102, 104, 120, 134, 137, 149, 151, 162, 182, 186-187, 189, 199, 203
New York Cement Show 18
Newberry, Spencer B. 17
Newton, Ernest, architect 13-14
Noel, Miriam xx, 9
Nordenskjold, Gustaf A.N. 158
Notre-Dame du Raincy, Paris 16
North Africa 145
Nute, Kevin 2, 87

Oak Park, Illinois vii, 15, 106, 149, 164
and Schindler 100-101
Oasis Hotel, Palm Springs 81
Ocotillo, camp/office 190
Oles, James 125
Olive Hill, Los Feliz, Los Angeles vii, ix, 8, 9, 41, 43-45, 102, 152-157, 163, 165, 180
Olmsted, Frederick Law, Jr. 20, 81
Olmsted, John Charles 81
On and By Wright 123
original action 2
ornament 24, 45, 122, 137, 145, 154, 182, 194, 204
Atterbury 19
FLW 2, 47, 85, 114, 134-135, 188, 192
— Bogk house 133
— German warehouse 132
— Imperial Hotel 103
— Midway Gardens 110, 112
— Unity Temple 15-16, 106
Gill 143, 199, 200-202
Greenough 31
Griffins' 127

Lloyd Wright 175
Loos 151-152
Morrill 22
Schindler 101, 176-178
Trost 129
Osborne, J.G., architect 13
Oud, J.J.P. 182
Wright School 190
Owls' Club, Tucson xi, 127, **129**

Palenque, Mexico 125
Palliser, Charles 17
Palos Verdes Estates, Los Angeles 180
Panama Canal 18
Panama-Pacific International Exposition 7, 148, 149
Pancho Villa 7
Paris Universal Exhibition, 1878 14
Parmentier, Fernand, architect 184
Pasadena, Calif. viii, 9, 32, 34, 36, 65, 94, 161, 204
Patterson, Terry 114
Perkins, Dwight Heald and Lucy 7, 55, 56-57, 59, 85-86, 191
and Mahony, Marion 55
Perret, Auguste, architect 16, 93-94
Persian faience 80, 193
Perth, Western Australia 63
Pfeiffer, Bruce Brooks xv
Pfeiffer Chapel 2
Phi Gamma Delta house 9
Philadelphia, Pennsylvania 87, 151
Pholiota (Griffin house) x, 180
Pinnell, Patrick 123
Ponce de leon Hotel 15
Pond, Allen B. and Irving K., architects 86
Port Chester, New York 13
Portland cement 11, 13, 23-24, 47, 73, 83, 147
Chicago-AA Portland Cement 132
Edison Portland Cement Company 14
Portland Cement Association xi, xvi, xvii
Proceedings 17
Practical Concrete-Block Making ... 17
Practical treatise on Coignet-Bèton and other artificial stone, 1871 13
Prairie School 191, 197n41
pre-Columbia 88, 119, 120-125, 136, 148, 154, 190
precedent 2, 22, 48-49, 99, 123-125, 130-133, 143
diversity of 191, 193

Greco-Roman 119, 124
progress before 87
Spanish 129
Price house (Gill) 6
primitive 22, 109, 119, 125, 132-136, 144, 190
progressivism 3
Purcell, William Gray xvi-xvii, 7, 55, 70, 191
and the Griffins 64-66
and the Taylors 58-59
pure design 85-88

Racine, Wisconsin xi, 100, 101
see also Hardy, Monolith, Johnson Wax
Radford, William A. 17
Architecture Co. 17
Ransome, Ernest L. 5, 11, 23
Unit System 14
Raymond, Antonin 172n71
Rebori, A. (Andy) N., architect 89-90, 104-105, 203-205
Red Scare 184
regionalism xxii, 4, 24, 32, 114, 143, 145-159, 165, 182, 187-189, 191
Rice, H.H. 17
Richland Center, Wisconsin xii, xx, 7, 14, 88, 130-131
Riis, Jacob 159
Riverside, Illinois 110
Robinson, Harry F. 7, 191
Roebling, John 137
Rogers, Gamble, architect 191
Rohe, Mies van der 122
Roorbach, Eloise 147, 199, 201
Mark Hopkins Art Institute 199
Roosevelt, Theodore 157
Ross, Denman W., architect 87
Rousseau, Jean-Jacque 119, 190
Royal Australian Institute of Architect 25
Royal Institute of British Architects 63
Rubbo, Anna xv
Russell Sage Foundation, see Sage Foundation

Saches, Martin, house project 186
Sage Foundation 20
Homes Company viii, 20
San Marcos-in-the-Desert 105, 114
Saylor, David O. 11
Saylor, Henry 12

San Diego, California xii, 6, 7, 32, 121, 131, 144-145, 147, **148**-151, 158, 165, 168n4-n5, 177, 178, 201, 206n5-n6
   Historical Society xii
   Panama-Pacific International Exposition 7
Saperstein, Morris *see* Philip Lovell
Scandinavia 19
Scheu house 152
Schindler, Pauline Gibling 35, 182
   Neutra 182
   political left 43
   and kindergartens 41, 44
   and the Lovells 43
Schindler, Rudolf (or Rudolph) Michael xi, xxii, xxv, 3, 7, 8, 51, 65, 115n5, 161, 163, 180, 192
   Chase/Schindler house *see* Kings Road house
   El Pueblo Ribera Court xiii, 9, 177, **178, 179**
   Freeman house ix, 39, 163
   How house 178
   Lowes house 153, 170n40
   Kings Road house xiii, 151, 175, **176, 177**
   — tilt-up concrete walls 176
   Lovell house 43
   Monolith, Racine 100-102, **101**, 153, **166, 167**
   Martin T.P., house xiii, 165
   Schindler house *see* Kings Road house
   and Aline Barnsdall 35, 45, 102
   — house 126-127, 157, 159, 163, 165, 172n64
   — Residence A & B 167
   and FLW xxiii, 3
   and Irving Gill 164
   and the Griffins 65-67, 81
   and L. Sullivan 163
   and the southwest xiii, 182, 194
Schmidt, Henry 124
Schmidt, Richard 191
Schwarzer, Mitchell 150
Scott, Margaret Helen xvi
Scottsdale, Arizona xv, xvi, xx, 97
Scripps, Ellen B. 7
Scripps Institution 6
Scully, Vincent 4
Sears, Roebuck Company 17, 14, 113, 117n45
Seattle, Washington xv, xvi, 8

Opera House 131
Secrest, Meyrle 123
Sell, Henry Blackman 137
Sergeant, John 123
Sewaren Improvement Company viii
Sewaren, New Jersey viii
Shaw, Howard, architect 191
Shaw, Richard Norman, architect 13-14
   Sketches for cottages and other buildings . . . cement slab system of W.H. Lascelles 13
Shaw and Lockington, architects 11
Shelly, Mary 3
Silsbee, Joseph Lyman, architect xx, 130, 149, 184
Silver, Joel xv
Simonds, Ossian Cole, landscape architect 159
simplicity 130, 176, 189
   Gill 143, 145, 150, 164, 176, 189, 199, 201
   Griffin 127
   and FLW 145
   — on Gill 151
Sinclair, Harry F. 185
Siry, Joseph 2
Smith, Bertha H. 147
Smith, Kathryn xv, 2, 123
Smith, William E. 8, 197n29
socialist 3, 41, 65, 184
South Dakota 158
Souvenir of the Shaw Block Company . . . made of cement concrete 17
Spain/Spanish 32, 93, 129, 149, 152, 156, 200
   colonial architecture 127, 146, 154-155, 159, 161, 192, 205
Spencer, Robert, architect 85, 191
Spider Web System 14
Sprague, Paul E. xv, 1-2
Spring Green, Wisconsin xx, 9, 10, 35, 66, 97n29, 101, 110, 160, 180, 181, 187, 197n29
   Women's Building xi, **111**
St. Augustine, Florida 15
St. Jean de Montmartre, Paris 15
Standardized Housing Corporation 21
Stanford University Museum 14
Stark, Paul x
Starr, Kevin 183
Staten Island, New York 12-13
Steiner house 152

Stewart house 153, 165
Stickley, Gustav 147
Storer, John 9, 31, 35, 39, 40, 161
　concrete block face ix, x, xii, 88, 132
　house viii-x, xii, 34-35, **36**, **37**, **38**, 44, 45, **46**, 82, 85, 90, 93, 153, 164, 187, 192
　construction problems 92
Stowe, Francis Ernest 7, 8, 54-60, 64, 70
　visit to U.S. 65
　visit to U.S. 1919 64-67, 81
　and the Griffins 65-67, 81, 180-181
Stuttgart, Germany 181
Suburban and rural architecture. Brick, stone, concrete and fireproof, 1867 13
Sullivan, Louis, architect xx, 7, 61, 86, 110, 120, 126, 143, 150-152, 161, 163, 188, 191, 192
　Cold Storage building 131
　death 9
　"form follows function" 31
　Lieber Meister 137
　Millet 86
　Tokyo 102-103, 108
　and Griffin 63, 67
　and Canberra 63
　and the Taylors 55-59
　and Trost 127
Sulman, John, architect 60
Supreme Court, U.S. 119, 137
Sweeney, Robert L. xi, xv, 2, 48, 80
Sydney, New South Wales 74, 179

Taliesin 85
Taliesin, Spring Green 2, 10, 20, 35, 97n29, 187
Taliesin Fellowship 20
Taliesin West, Scottsdale 20, 25
Tallmadge, Thomas, architect 86, 191
Taos, New Mexico xiii, 7, 165
　pueblo 154, 173, 182, 193
Taylor, George Augustine and Florence xvi, 53-67, 69
　Peace City 1922 180
　visit to U.S. 1914 7, 8
　visit to U.S. 1922 180
　and the Griffins 65-67, 81, 180-181
Tea Pot Dome, Wyoming, scandal 185
Teat, Katherine house 145
Temple of the Sun 125
Temple-Poole, George 63

terra cotta/clay x, xi, xii, xiv, 8, 90, 14, 15, 47, 61, 73, 79, 90, 107, 112
　fire-glazed 80
　Imperial Hotel 126
　no structural value 100
　Winslow house 110
　see brick
　see Natco
textile block viii, 9, 43, 74, 80, **82**, 83-85, **84**, 94, 102, 105, 146, 160, 164, 180, 182, 204-206
　aesthetics **84**, 192-193
　Arizona Biltmore 90
　problems 98n51-n52, 113-114
　textile tiles 10, 31, 39, 47, 73, 184, 185
　and Irving Gill 147
　and Natco 80
　and Neutra's Lovell house 183
　as a failure 92
　see also concrete block
Tilt-Up Concrete Association
　Irving Gill Distinguished Architect Award 169n26
Timkin, Henry F., house 201
Toker, Franklin 2
Tokyo Japan 7, 43, 81, 101-103, 186
　see also Imperial Hotel
Tomlinson, Webster, architect xx, 86
Torrance, California 81, 175
Tourcoing, France 23
Trost, Gustave **129**
Trost, Henry C. xi, **129**
Trost and Trost, Architects **129**
Truscon Corporation 14
Tsuchiura, Kamecki, architect 197n29
Tucson, Arizona xi, 127-**129**
Turnbull, Jeff, architect xv
Turner, C.A.P. construction system 6, 14, 140n40
Twombly, Robert 123

United Kingdom/Britain 19, 55, 58
U.S. Army, Mexico 7
　Wounded Knee Creek 158
U.S. Bureau of Ethnology 158
United States Shoe Machinery Company 5
Unity Temple vii, 2, 15-16, 18, 73, 88, 92, 106, 113, 126, 163-164, **186**, 192
Unity Temple, FLW and Architecture for Liberal Religion 48
Universal Church, Oak Park 15

see also Unity Temple
Universal Portland Cement Company 24
   competition 17
   U.S. Steel Corporation subsidiary 18
University of Chicago 159
University of New Mexico
   Science Hall xi, 127, **128**
   see Byrne, Griffin
University of South Australia 25
University of Southern California 34
University of Wisconsin xx, 148
Uxmal, Yucatan, Mexico
   Mayan buildings xi, xii, **124**

Van Bergen, John, architect
   Wright School kit houses 113, 117n45
Van Zanten, David 123
Veracruz, Mexico 7
vernacular 119, 122
   Algerian 152
Vernon, Christopher xv
Viennese secession 2
Viollet-le-Duc, E.-E., architect 2, 15, 85, 86, 120, 141

Wagner, Betty xv
Wagner, Otto, architect 63, 165, 181
Wales 85
Wallatzen, Andrew, architect 191
Ward, George A. 12
Ward, William E., engineer 13
   Ward's Folly/Ward's Castle 13
Wasmuth publisher 6, 56, 152, 161, 189
Watson, Vernon S., architect 191
Wendingen xix
West Hollywood, xiii, 7, 150-151, 175-176
Western Architect 199
Whipple, Harvey 18
White, Charles E., architect 191
Whitlesey, Charles F., architect 184
Whitman, Walt 173, 143-144, 152
Wie baut Amerika? (How does America build?) 181
Wight, Peter B. 86
Wilde, Oscar 57
   "All art is quite useless" xxiii, 152
Wilkes, Paul 17
Wilkinson, William B. 11
Willey house 163
Willis, George, architect 191
Winslow, William Herman, house 6
Wittekind, Henry 17

working class housing viii
Women's Building and Neighborhood Club 110, **111**
World's Columbian Exhibition 151
Wounded Knee Creek 158
Wright, Anna Lloyd xx, 9, 31, 85
Wright, Catherine Tobin 6
   kindergarten 121
Wright, Eric Lloyd xv, xvi, 175, 207
Wright, Frank (Lincoln) Lloyd
   biography and portrait xx, xix
   critical events 5-10
Wright, Frank Lloyd, Jr. *see* Wright, Lloyd
Wright, Gwendolyn 123
Wright, John Kenneth (Lloyd) 6, 7, 148
Wright, Lloyd (Frank Lloyd Wright, Jr.) vii, viii, xii, xiii, xiii, xvi, xxii, 6, 18, 35, 39, 47, 51, 94, 120, 140n39, 164, 165, 175, 170n42, 171n58, 172n77, 175, 177
   architect's house 159
   architect's license 149
   Arizona Biltmore Hotel 90
   Bollman house 80, 130, 159
   block construction 90, 108
   collection xvi, 207
   father's LA troubles 184-185, 206
   Lowes, G.P., house 153
   Irving Gill 150, 165, 188
   — "cube house" **150**
   knit-block 73, 79-80, 193
   Los Angeles 162, 184, 197
   Millard house viii, 33, 85, 163
   Paramount studio 149
   Samuel-Navarro house 133
   San Diego 81, 145-146
   Sowden house 96n18
   and Aline Barnsdall 25n2, 157, 163
   and Irving Gill 145-152
   and the Griffins 181-182
   and Schindler 66
   and the southwest 154-155, 159
Wright, William Carey 18, 31, 85
Wright in Hollywood 2
Wright School xii, 113, 191
"Wright versus America" 189

Yellowstone National Park 160
Yokohama, Japan 8
Yosemite National Park 159, 160, 175

Zuni Creation Myths 158

Made in the USA
Monee, IL
03 May 2026

49437965R00144